A Student's Guide to Natural Science / Psychology

学科入门指南
自然科学·心理学

[美]史蒂芬·M.巴尔 [美]丹尼尔·N.罗宾逊 著
刘慧梅 潘寅儿 译

浙江大学出版社
ZHEJIANG UNIVERSITY PRESS

目 录

自然科学入门指南

引 言	3
科学的诞生	5
科学的第二次诞生	11
科学、宗教和亚里士多德	14
科学革命	23
科学方法	23
从哥白尼到牛顿	28
数学发挥新作用	31
牛顿物理学	35
力和场	39
20 世纪的物理学革命	43
相对论	43
相对论有多大的"革命性"？	47
量子革命	51
对称性的作用	56
"数学无理由的有效性"	59

心理学入门指南

引 言	63
哲学心理学：学科的创立	66
古希腊时期（起源）	66
苏格拉底和柏拉图	69
亚里士多德及其自然观	73
亚里士多德之后的哲学心理学	77
作为自然科学的心理学	80
达尔文进化论	83
神经心理学和神经学的进步	85
行为主义	86
神经心理学和认知神经科学	91
弗洛伊德和精神分析学	95
社会心理学	99
人类发展：道德和文明	101
结　语：永恒的问题	104

CONTENTS

A Student's Guide to Natural Science

Introduction	109
The Birth of Science	111
The Second Birth of Science	119
Science, Religion, and Aristotle	124
The Scientific Revolution	136
The Scientific Method	136
From Copernicus to Newton	143
Mathematics in a New Role	148
Newtonian Physics	153
Forces and Fields	159
The Twentieth-Century Revolutions in Physics	164
The Theory of Relativity	164
How "Revolutionary" was Relativity?	170
The Quantum Revolution	175
The Role of Symmetry	181
"The Unreasonable Effectiveness of Mathematics"	186

A Student's Guide to Psychology

Introduction	191
Philosophical Psychology Inventing the Subject	195
Ancient Greece (Again)	195
Socrates and Plato	198
Aristotle and the Naturalistic Perspective	204
Philosophical Psychology after Aristotle	210
Psychology as Science	215
Darwinian Evolutionary Theory	219
Advances in Neurophysiology and Neurology	221
Behaviorism	224
Neuropsychology and Cognitive Neuroscience	230
Freud and Depth Psychology	236
The Social Context	241
Human Development: Moral and Civic	244
Abiding Issues: An Epilogue	249

自然科学入门指南

引 言

自然科学包括物理学、天文学、几何学、化学和生物学，通常被认为是与人文科学，如人类学、社会学和语言学等相对的科学。当然，两者之间还是有重合的部分。把人类作为生物有机体来研究的学科，就既属于自然科学又属于人文科学。

在这样一本导读中，要均衡地涉及自然科学的所有分支是不可能的。因此，我选择着重讲物理学，略带讲一些天文学。做这样的选择有几个原因：首先，这些领域的突破促成了科学革命，并且开创了现代科学时代。第二，物理学是自然科学最根本的一个分支，因为物理定律统领着所有其他分支。自然科学经常用一种自下而上的方法看待事物，复杂系统的行为是以构成成分间的相互作用来解释的。自然科学研究物质最基本的构成以及他们之间相互作用的分支就是物理学。第三，我们可以说，物理学和天文学的发展，与达尔文的进化论一样，对于哲学思想有着最深远的影响。最后一点，我本身就是个物理学家。

亚洲、非洲和美洲的每一个伟大的古代文明都曾进行科学研究。但是，就如人们常说的，科学是从古希腊人以及他们的巴比伦和埃及的前辈那里，通过伊斯兰世界传入欧洲的。这种说法是有据可依的。17世纪在欧洲爆发的科学革命源自古希腊人的成就，而所有的现代科学都起源于这次科学革命。古代世界其他地区的科学发展，虽然其本身也相当令人震撼，但是它们对于西方的科学革命没有贡献或是贡献甚微，因此也就没有持久的影响力。(当然也有例外，比如，数字零的概念是印度人首先提出的，然后再由阿拉伯国家传入欧洲。)从16世纪到19世纪，几乎所有的科学进步都无一例外地来自欧洲。直到20世纪，科学才成为一项全球性的事业。

科学的诞生

约在基督诞生500年前,西方自然科学诞生于希腊。它是由两个伟大的观点碰撞产生的。第一个是我们可以系统地运用理性来扩大我们对于现实的理解。在这一方面,可以说是希腊人发明了"理论"。举例来说,虽说自从有创作的时候起就有了文学,有人的地方就有了政治,政治理论和文学理论却是希腊人的发明。逻辑研究和数学公理的发展也是如此。希腊最早的哲学家之一赫拉克利特(前540—前480)曾说,世界处在永恒的混乱之中,但理性或逻各斯潜藏于所有的变化之下。

第二个伟大的观点是物理世界的事件可以有合乎自然规则的解释,而不是超自然的或是纯神学的解释。这种提法的先驱是米利都学派的泰勒斯(前625—前546)。据说,他曾提出地球是漂浮在水上的,以此来解释地震。他最著名的猜测是水是万物的本源。所以泰勒斯或许是寻找构成万物的基本元素(或在他的例子中,唯一元素)的第一人。其他人提出了不同的元素,最终列表上的元素达到四个:火、水、土和空气。

直至今日,人们还在继续寻找构成世界的真正根本或基本的成分。1869年,门捷列夫公布了他的化学元素周期表(当时有63种元素)。后来,化学家们证实之前发现的原子是由亚原子粒子构成的,这些粒子是科学的一个分支——基本粒子物理学所研究的对象。今天人们推测这些粒子并不是最基本的,它们本身是"超弦"的表现形式。如果证明现在这种推测是正确的，那么就证实了泰勒斯的直觉，自然界只有一种真正根本的组成成分。事实上，正如我们看到的，随着每一次现代科学的重大进步，这种理论统一化和简单化的梦想已经在逐步地实现。

"原子"的概念是所有古希腊科学观点中最引人瞩目也最具远见的，它首先由阿夫季拉的留基伯（前5世纪）和德谟克利特（约前460—前370）提出。诺贝尔奖得主理查德·费曼在他著名的《物理学讲义》中写道：

> 如果在某次大灾变中，所有的科学知识都要被毁掉，只有一句话可以传到下一代，怎样的陈述能用最少的词包含最多的信息呢？我认为那就是原子的假设……万物皆由原子构成——微小的粒子处在永恒的运动中，当它们之间存在一定距离的时候就相互吸引，但要被挤压成为一体的时候又相互排斥。

当然，从现代意义上来说，留基伯和德谟克利特提出的初步的原子论不是一种科学理论。它无法被验证，也没有引发任何研究项目的开展，而是像大多数希腊自然科学那样停留在哲学思考的层面上。

阿基米德（约前287—前212），数学史上最伟大的人物之一，出生于西西里岛的叙拉古。他发明了计算曲面物面积和体积的方法，在17世纪，托里拆利、卡瓦列里、牛顿和莱布尼茨为创立微积分又进一步发展了这些方法。与大多数古希腊数学家不同，阿基米德对物理问题兴趣浓厚。他是第一个理解"重心"概念的人。他还创立了流体静力学领域，发现漂浮的物体会排开等同于它自

所有科学分支的初始阶段都包括简单的观察和分类。希腊大部分的自然科学都是由这类活动构成的,这一点不足为奇。有时候,它也会雄心勃勃地追寻起因和原理,但这些原理主要还是停留在哲学层面的。换句话说,这些原理没有成为现代意义上的科学定律。想一想亚里士多德的原理,"任何物体只有被另一物体推动时才会运动",这是对于因果的一个总体陈述。我们无法依此推测任何事,更不用说计算了。

身重量的液体,而浸在液体中的物体会排开等同于自身体积的液体。他用上面第二种原理解决了锡拉库扎的希罗王交给他的难题,即在不熔化皇冠的前提下,判断它是否由纯金打造。一天在公共澡堂洗澡的时候,他忽然想到了解决的办法,就光着身体跑到街上,还边跑边喊"尤里卡!"(希腊语:"发现了!")这是科学发现者不朽的呼喊啊!

阿基米德还是杠杆原理的发现者,他曾自豪地说"给我一个支点我就能撬动地球"。据说通过发明神奇巧妙的武器,如阿基米德之爪,还有用来引燃敌船的巨大的聚焦镜等,阿基米德帮助叙拉古在第二次布匿战争中抵御了罗马的围攻。普鲁塔克(生活于罗马时代的希腊作家)曾说:"(阿基米德)一直着迷于他熟识的一位女神,那就是他的几何学,因此而废寝忘食,无暇照顾自己……他经常被拽去澡堂洗澡,在火堆的灰烬中画几何图形,用沾满油的手指在身上划线,他总是处在一种狂喜的状态中,极度沉迷于他的科学。"在叙拉古被围困的时候,尽管罗马将军有令不能伤害这个伟大的几何学家,阿基米德还是被一个罗马士兵杀死了,当时他正在沙地里画几何图形,他最后一句话是"别踩坏了我的图!"

虽然希腊人在数学方面有众多成就，却没有将数学运用到他们对物理世界的研究中——天文学是一个主要的例外，这是一个有趣的现象。但我们不该为此感到讶异。世界是有序的，而不是在一片混乱当中，这也许是很显然的，但是只要看到地球上的事物和事件存在那么多的不规律性和偶然性，这种有序性是数学化的这个事实就远没有那么明显了。毕达哥拉斯（约前569—约前475）第一个提出数学是理解物理实在的根本，而非只是与某种理想境界相关这个观点。这种见解可能来自于他对音乐的研究，他发现和谐的音调是由长度相互之间成简单数学比例的弦发出的。不管怎样，毕达哥拉斯和他的后继者最终形成了这样一个观点——最深层次的现实是数学化的。事实上，亚里士多德将"万物皆数"的思想归因于毕达哥拉斯。这种说法可能看起来很极端，毫无疑问，亚里士多德也是这么认为，但是对于现代物理学来说它似乎意义深远而且颇有预见性。

宇宙的数学有序性在天体运动中最为明显。这与一些条件有关。首先，太阳系内的空间几乎是真空的，也就是说太阳系各种天体的运动不受摩擦的影响。第二，行星相互之间的引力与它们跟太阳之间的引力相比是很小的，这大大简化了它们的运动。换言之，在太阳系中，自然为我们提供了一种相对容易分析的动态系统。这对科学的出现至关重要。在经验科学中，能将特定的原因和结果区分开来是很重要的，这样它们就不会被外部的或者不相关的因素混淆或者干扰。这通常需要进行对照实验（例如，一些实验能够让人对只有单一变量的两个系统进行比较）。往往只有通过这种方法，人们才有机会观察到数据中有趣又重要的模式。但是,古希腊人没有想到要进行对照实验，或者说大部分希腊人没有想过进行任何类型的实验。因此，他们有太阳系可以观察是很幸运的。

首次将几何学应用于天文学似乎是受了毕达哥拉斯思想的启发。毕达哥拉斯自己曾提出地球是一个正圆的球体。后来，欧多克索斯（约前408—约前355）提出了一个模型，认为看似复杂的天体运动其实是它们沿着正圆轨道运动的结果。这些毕达哥拉斯的原理和理论有着一种数学美而且它们可以用简单的原因解释复杂的现象，这在科学史上硕果累累。但是它们还有不足之处。这些早期天文学家的数学方法缺乏一种重要的因素，即进行精确测量并将理论建立在这些测量结果之上。在这方面，喜帕恰斯（约前190—约前120）远远地超越了前人，将天文学转化为一门定量的预言性的学科。他极其精确地测定了这样一些量，例如，月球和地球之间的距离和地轴进动的速率（所谓

喜帕恰斯（前190—前120），被认为是古代最伟大的观测天文学家。我们对他的生平不甚了解，除了知道他出生在尼赛尔——位于今天的土耳其，并且在罗得岛上度过了一生中大部分的时光。让他异于前人的是，他将精确测量运用到天文学的地心模型中。他不仅做了大量的测量，还用到了可追溯至公元前8世纪的巴比伦人庞大的天文学记录。跨度如此之长的数据使他能以前所未有的精度来计算某些量。

喜帕恰斯创造了三角函数表，这极大地促进了天文学的计算，他还发明或是改进了天文观测设备。另外，他还编制了第一张星表，标明了大约1000颗星的位置。虽然他研究很多问题，如测量地球到月球的距离，但他是以发现"分点岁差"现象并正确地将之归因于地球自转轴的摆动而闻名的。后来牛顿说明，这种摆动是太阳和月亮在地球赤道区隆起带上的引力扭矩造成的。

的分点岁差就是他发现的一个现象）。在喜帕恰斯之后，因为托勒密（约85—约165）的成就古希腊天文学的发展到达巅峰，他提出的"地心说"在之后的15个世纪被广泛接受。

  希腊人对于数学的兴趣是一把双刃剑。一方面，它对科学的发展有巨大的益处。希腊影响最久远的科学遗产就在于他们的数学和数理天文学。另一方面，这种对于数学的兴趣反映了一种贬低现象世界而更专注于理念世界的倾向（这在柏拉图身上表现得非常明显）。

  在亚里士多德身上，我们看到了一种几乎相反的倾向。亚里士多德对每种现象都兴趣十足，（与许多他的后继者不同）他进行了广泛的实证研究，尤其是在生物学领域。毫无疑问，亚里士多德是古代世界最伟大的生物学家之一。但他在物理学方面的遗产可以有多种解读，事实上，从整体上来说可能是负面的。作为一名物理学家，亚里士多德相对失败有几个原因。首先，已经谈到的一个事实是很难对地球上的现象进行分类。在其他很多难题之中，有一个是地球现象涉及很大的摩擦力，这导致亚里士多德在力与运动的关系上彻底被误导了。第二，亚里士多德不懂得数学的真正本质，也意识不到它深刻的重要性；他的才华在于其他方面。第三，亚里士多德研究物理科学的方法是哲学式的。在他看来，物理学概念和纯哲学概念没有明确的分界线。若不是亚里士多德哲学的智慧与深度，这种观点本不会给后来的思想带来这么多的问题——这些问题将在以下部分仔细讨论。

科学的第二次诞生

那些别有用心的宗教反对者有时会说，基督教的兴起终结了科学进步的第一个伟大时代。这种说法是站不住脚的。的确，我们能在教父所写的文章中找到反对自然研究的论述，而且科学并不是早期基督徒关心的重点。但是，我们发现早期的基督徒与他们同时代的非基督徒对于科学的态度是一样的。而且，事实是在基督徒成为重要的人口组成部分或知识分子力量之前，古代科学的辉煌时代早已过去。希腊数学的黄金时代在基督诞生200年前就结束了。（比如，希腊伟大的数学家阿基米德、埃拉托斯特尼和佩尔格的阿波罗尼奥斯分别于公元前212年、前194年和前190年去世。）古希腊科学中，只有一小部分伟人生活在基督诞生以后，特别引人注意的是大约于165年去世的天文学家托勒密，还有284年左右去世的数学家丢番图。那时，基督教还是一个很小的遭迫害的教派。

众所周知，数学和科学令人瞩目的复兴始于9世纪的伊斯兰世界。在疆域从北非一直延伸到中亚的阿巴斯伊斯兰王权的统治下，学者们能够借鉴巴比伦人、印度人和希腊人的遗产。人们记得穆斯林对于科学的贡献是因为许多源自阿拉伯语的科学术语，如化学（曾经被称为"炼金术"）中的酒精和碱，数学中的代数、算法和数字零，还有天文学中的太初历、方位角、天穹以及亮星的名字如大陵五、毕宿五、参宿四、参宿七、织女星。但是，几个世纪之后，穆斯林科学的光辉就开始退去了。伊斯兰的神学机构对于纯理论的希腊思想倾向于持漠视或敌对的态度，因此，科学在穆斯林世界没能像之后在欧洲大学里那样获得一种建制化的地位。

科学的第二次诞生真正起源于西方拉丁世界。11 世纪，当欧洲开始从野蛮人入侵造成的经济和文化崩溃中恢复过来的时候，欧洲学者主要是通过与阿拉伯世界的接触重新意识到了古希腊科学的伟大成就。这种意识引起了对古希腊学者的作品无尽的好奇以及需求，这进而又掀起了一场将从西班牙获得的阿拉伯语作品译成拉丁语或是直接将从拜占庭获得的希腊语作品译成拉丁语的翻译热。大学最早成立于中世纪的欧洲，建立大学的部分原因就是为了让人们能在此学习这些刚刚被重拾回来的知识。神职人员和平民对希腊科学（当时的称法是"自然哲学"）都抱有强烈的兴趣。实际上，在中世纪的大学，学习自然哲学是学习神学的前提条件。（这和今天物理是高中的必修课程有些相似。）

有很长一段时间，现代学者都习以为常地将中世纪的科学置之一旁，因为它缺乏创造力或是真正的科学精神，也因为它与后来的科学进步几乎无关。但是，像皮耶尔·迪昂和 A. C. 克龙比这样的学者成功地挑战了这种普遍的观点。他们证明中世纪的科学远比之前所认为的重要，而且如同奴隶一般跟随着亚里士多德的苦行僧式的学者已经有太多了。虽然中世纪的"自然哲学家"的批判法不是基于实验而是基于逻辑推理，从某种程度上来说是基于我们今天所说的"思维实验"，但他们非常清楚亚里士多德思想中的不足之处，并用一种谨慎的批判态度来看待他。此外，中世纪的人向发展一门运动科学迈出了试探性的脚步。人们第一次理解了"等加速"（uniform acceleration）（或者用他们奇怪的术语来说"统一的不一致的"运动）这个关键的概念；第一次提出了"动力"（impetus）的概念（"动量"momentum 概念的前身）；发明了图表以进行数学函数和运动的推理；而且第一次提出了运动的数学定律。一些历史学家如迪昂、克龙比和更为近代的斯坦利·杰基甚至宣称，这些观点直接影响了伽利略和其他科学革命奠基人的思想（虽然关于这种影响的程度是有争议的，而且这个问题还远未解决）。

尽管如此，中世纪欧洲科学的复兴确实以某种方式奠定了之后科学革命的基础。正如著名的科学历史学家爱德华·格兰特所说，它将科学"制度化"了。在古代世界和阿拉伯世界中，科学就像艺术，依赖于一些个人的赞助，这些人富有或是有权力同时又恰好对科学感兴趣。因此，科学是件碰运气的事，受制于风云变幻的政治经济形势。相反，在中世纪的大学里，首次出现了一代接着一代不断研究科学问题的稳定的学者团体。也就是说，一个科学团体形成了。到中世纪末，欧洲大约有100所大学，其毕业生数以万计。这就创造了对科学感兴趣且有文化的一个重要的公众群体，他们愿意花钱接受科学教育或者购买科学的相关书籍，在这些人的队伍中就会出现科学人才。

如果没有中世纪大学所创造的科学团体和科学公众群体，科学革命就没有肥沃的土壤得以萌发。

科学、宗教和亚里士多德

之前的讨论引发了两个非常有意思的难题。为什么科学革命发生在欧洲,以及宗教在这次革命中扮演了怎样的角色。有一种观点,因为我们听得太多了,已经成了一种陈词滥调,那就是基督教与科学为敌而且想把科学扼杀在摇篮里。这种敌意在伽利略事件中已经展露无疑。但是,现在学者们不再把这种观点当真了。教会权势集团对科学抱有难以平息的仇恨这种观点是毫无根据的,是为反宗教和反教权目的服务的。

事实上,教会一直尊重科学研究,即使在伽利略事件的当时也是如此。我们已经知道虽然古希腊科学本质上是自然主义的而且从起源上来说是异教的,中世纪的教会还是愿意接受它;我们也知道这些科学在中世纪大学的课程中占有非常重要的地位,而大学这样的机构主要是在教会的支持下建立起来的,接受了教会大量的赞助和保护,教师中的大部分也都是神职人员。实际上,中世纪大部分科学家都是牧师,如尼古拉斯·奥雷姆(约1323—1382),他是利雪主教,也是一位杰出的数学家和物理学家。这种牧师参与科学研究的传统已延续到了今天。事实上,从19世纪一直到20世纪,天主教的牧师作出了大量重要的科学贡献。

教会对"自然哲学"持支持的态度主要是大阿尔伯特和他的学生托马斯·阿奎那努力的结果,前者将希腊科学引进了中世纪的大学。这两个人都相信将信仰和理性融合的可能性以及重要性，并在亚里士多德的哲学中找到了实现这种融合的概念工具。这对自然哲学的研究起到了巨大的推动作用，因此也为科学革命铺平了道路。但是，教会欣然接受亚里士多德也造成了负面结果，因为这正好帮助强化了一种错误的研究自然科学的方法。

尼古拉斯·奥雷姆（约1323—1382），出生于诺曼底的卡昂附近。他在巴黎大学取得神学博士学位，之后成为法国国王的顾问和牧师，最后就任利雪区主教。他是一位博学广识的天才，在数学、物理学、音乐和心理学方面都作出了开创性的贡献。他是经院哲学的重要人物，并被很多人认为是中世纪最伟大的经济学家。在数学方面，他是第一个论述分数甚至无理数的人，他对无穷级数的研究证明了调和级数（$1+1/2+1/3+1/4+\cdots$）无限大。他也是第一个用图表表示两个变量之间函数关系的人，并将论述延伸到三维图表，因而在笛卡尔之前3个世纪就预言了其解析几何学的一些关键思想。他第一次用图表法证明了"莫顿定理"，该定理可以计算匀加速运动的物体通过的距离。这些研究很可能间接地帮助伽利略发现了落体定律。

此外，奥雷姆证明了如果假设是地球围绕天空转而不是反过来的话，可以更圆满地解释很多现象。他用来驳斥普遍物理学反对意见的分析比后来哥白尼和伽利略提出的日心说更加出色，尤其是将运动分解为水平和竖直两个部分。他是第一个推断光在通过不同密度的空气时发生折射，因而导致地平线附近天体的位置失真的人。他认为存在着一个无限的宇宙，这与标准的亚里士多德的观点相反。总而言之，他必然是数学和物理学史上最具独创性的思想家之一，也是科学革命的一位重要先驱。

亚里士多德的哲学体系经托马斯·阿奎那的改造和基督教化以后成为了光彩夺目、令人赞叹的智慧结晶。我们可以认为它是中世纪的"大统一理论"或"万有理论",只是它不仅包括了自然哲学,还包括了形而上学、天文学、伦理学及很多其他学科,甚至包括了天启教的真理(只要它们能被人类的理性所理解)。这种亚里士多德—托马斯的哲学有很大的影响力,而且实际上可以说是作为一种活的传统保留了下来。但是,尽管亚里士多德的思想阐明了许多玄学和道德方面的问题,当涉及物理学和天文学的时候,他的思想中还是有很多是极其错误而且具有误导性的。

责怪中世纪的人采纳了亚里士多德的物理学是愚蠢的,因为他们别无他选,这是他们从希腊人那里继承下来的科学。就像错误理论也能做到的那样,在一段时间内,它的确起到了为理论讨论和理论分析提供框架的功效。它也为用自然主义理论解释物理现象提供了例证,这一点意义不小。但是,它也把科学带进了一条死胡同,后来费了很大劲才从中逃脱。从神学有助于延长亚里士多德自然哲学的统治地位这方面来说,它起到的作用并无裨益。

亚里士多德主义的统治地位有助于解释 1633 年教会对于伽利略和日心说的谴责。但是,它仅仅是一次复杂事件中的一个因素。给伽利略定罪的其他原因包括职业上的竞争,伽利略善于树敌的天性,还有最重要的是时局的动荡。那是一个宗教关系极度紧张的时代,欧洲正被以天主教和新教的斗争为开端的三十年战争弄得分崩离析。作为抵御新教挑战所做的努力之一,天主教教会在特伦托会议(1545—1563)颁布了关于圣经解读的一套规定,目的就是防止激进的宗教革命。虽然其本身是合理的，但这些规定最后却错误地用在了伽利略身上，他很不明智地让自己被他的敌人拖入了一场圣经和宗教的辩论之中。

伽利略·伽利莱（1564—1642），意大利比萨人。从1581年开始在比萨大学学习，并于1589年获得比萨大学教授职位。由于与亚里士多德派学者的冲突，两年后决定去帕多瓦大学任教。1609年，在得知望远镜发明之后，他也设计了自己的望远镜并开始用它研究天空。他发现了太阳黑子、木星的卫星、月球的环形山和金星的相位，这些发现破坏了亚里士多德的科学，也驳斥了托勒密体系，使他成了名人。他对于哥白尼日心说体系的拥护招来了反对，1616年，罗马宗教裁判所颁布了一条禁令，禁止他以任何方式维护哥白尼主义。同时，所有宣扬哥白尼主义为真理而不仅仅是一种"假设"（所谓"假设"是指用来简化数学计算的一种方法）的书都被列为禁书。1623年，哥白尼的一个朋友兼保护者马菲奥·巴贝里尼被选为红衣主教。因为没有意识到1616年的禁令，所以他没有反对伽利略把哥白尼主义作为一种"假说"来维护。1630年，伽利略进而出版了《关于两种世界体系的对话》(*Dialogue Concerning the Two Chief World Systems*) 一书，在书中，他不仅将哥白尼主义作为真理来捍卫，而且似乎激烈地讽刺了主教的哲学观点。被一个曾经保护过的人背叛，这触怒了主教，被遗忘的1616年禁令被人从文件堆中挖了出来。1633年，伽利略被迫公开宣布放弃哥白尼主义并被判终身囚禁在家。他在他佛罗伦萨的别墅中服刑，他可以接待拜访者，出版关于其他科学问题的书。1638年，他出版了《关于两种新科学的对话》(*Dialogues Concerning Two New Sciences*)，书中陈述了他的物理学发现，这是他对科学最伟大的贡献。

对伽利略的判罪虽然是个弥天大错，但并不是教会当权派仇视科学的结果，也没有反映出教会在科学问题上的顽固武断。1616年伽利略第一次遭罗马宗教裁判所审查时，主事红衣主教贝拉明的话很值得我们回味：

> 如果得到证明（太阳的确是静止的，而地球是运动的），那么我们在解读《圣经》篇章时就得采取谨慎态度了，因为这些篇章似乎说的正好相反，我们宁可承认自己不理解它们，也不能宣称那些已被证明为真的东西是假的。

贝拉明继续说，他对存在这样的证据持非常怀疑的态度，还说如果有怀疑，那么人们就必须保留对《圣经》的传统解释。事实上，在伽利略的时代，确实还不存在日心说的证据，但是对于那些能看到的人来说，的确有明显的迹象表明日心说是正确的。

不管怎样，如果我们看看教会与科学800年关系的记录，很难不把伽利略事件仅当成一次误差。一些学者已经提出，基督教信仰在使科学革命成为可能这一点上功不可没，而远非将宗教视为现代科学出现的拦路虎。关于这点有很多证据。例如，所有科学的一个根本观点是存在着一种自然规则，即不仅世界是有序的，而且事实上这是一个"自然的"世界。我们已经知道这个观点源于异教的希腊哲学家，但犹太教和基督教也帮助发扬了这种思维方式。但是，在最初的异教徒信仰中，世界充满了超自然的神秘力量，住着无数的神——战争之神、海洋和土地之神、性和生育女神等，但犹太教徒和基督徒认为只有一个神在自然界及其现象和力量中是找不到的，只有在自然界外才能找到，他是真正的自然的主人。这样，圣经信仰就把世界世俗化和客观化了。借用马克斯·韦伯的话说，它使世界摆脱了幻想。

又如，学者们告诉我们，《创世纪》(Genesis) 这本书经常被视为原始神话创作的范例，而实际上，创作这本书的部分原因是为了驳斥异教徒的超自然主义和迷信。当《创世纪》中说，太阳和月亮只是上帝在天空中安的灯，用来照亮白天和黑夜，

尼尔斯·斯坦森（1638—1686，也被叫做斯旦诺），对解剖学、几何学、古生物学和晶体学作出了重要贡献。他出生于丹麦，就读于哥本哈根大学药学系。二十几岁的时候，他就已经被认为是欧洲最出色的解剖学家之一。他的解剖学研究大大扩充了人们对于腺体淋巴系统的认识。（斯坦森管、斯坦森孔和斯坦森腺都是以他命名的。）他在心脏和肌肉结构、大脑解剖学和胚胎学方面也做了重要的研究。在巴黎和芬兰呆了一段时间后，他去了佛罗伦萨并加入了西芒托学院。该学院由托斯卡纳大公斐迪南·德·美第奇创立，旨在依照伽利略的传统进行实验研究，而伽利略的一些学生也是该学院的成员。在解剖从里窝那（意大利西部港市）捕获的大白鲨的头部时，斯坦森发现鲨鱼的牙齿和在马耳他大量发现的"舌石"惊人地相似。于是他开始进行调查，也因为这些调查，他被认为是化石科学研究的始祖和地质学分支之一地层学的奠基人。其关于地理地层是如何形成的观点，在解开地球历史之谜以及最终揭示地球高龄的过程中扮演了重要角色。此外，他还发现了晶体学的基本定律，即一块特定矿石的所有角都是一样的（"斯旦诺定律"）。1667年斯坦森转信天主教，1675年被任命为牧师，两年后成为主教。作为一名主教，他奉行严酷的禁欲主义，还是穷人热心的捍卫者。1988年，教皇约翰·保罗二世为其行宣福礼。

其实它是在攻击异教宗教对太阳和月亮的崇拜。此外，当书中说人是依上帝之形创造出来管理动物的，它是在攻击崇拜并且服从动物或者化为动物形体的神的异教主义。

中世纪的基督徒非常欣然地接受了物理世界的自然主义观点，根据爱德华·格兰特的说法，早在 12 世纪，哲学家和神学家把宇宙称为一台机器是司空见惯的事。（当然，犹太教和基督教也相信"奇迹"miracle 的可能性。从拉丁语 mirari 意为"感到惊讶"转变而来的 miracle，恰恰是因为非常罕见而与中世纪哲学家所说的"自然规律"相违背。它们所有的意义都来自这样一个事实：自然界存在着一种秩序，只有作为自然之主的上帝才能凌驾于其上。）

犹太教和基督教可能帮助科学革命形成的第二个观点是宇宙不仅仅是有序的还是"合法"的。公元 2 世纪的基督教作家米努修曾写道：

> 如果你进入一间屋子，发现里面所有的东西都精心装饰，整齐干净，呵护有加，你会认为有个主人在管理这个家，而且他本身要比这些好东西高级得多。所以同样的，在世界这个家中，当你在天空中、在地球上看到天意、秩序和法则，请相信一定有一个宇宙的王者或者主人，比繁星本身和整个世界中各色各样的事物更美。

这种对于"法"（law）的强调可以追溯到希伯来经文，它的前五本书被犹太人称为"律法"或"法"。当然，以色列人认为上帝是他们的法律制定者，同时也是宇宙本身的立法者。上帝在《耶米利书》中说："如果我与日夜之间没有圣约，没有给天地立法，那么我也不会让雅各和我的仆人大卫有后代。"《诗篇》第 48 章中说日、月、星辰和天空都遵守一条神赐的不会消亡的法则。古代犹太教的教士说:"上帝,虽然万福齐身,在创造世界的时候还是要请教律法。"因此,律法是存在于上帝心中的一条法律,宇宙本身就是根据这条法律创造出来的。(在犹太人看来,这种永恒先存法则的观点与犹太神性智慧的观点相关,这在《旧约》后面部分的书中有所体现,尤其是在《箴言篇》《智慧篇》《西拉书》和《巴鲁书》中。这种神性智慧在《新约》中成了"逻各斯",意为"道"或"理性"。换言之,在异教的希腊人和犹太人的思想中都形成了宇宙之下潜藏着理性这个观点。)

不是只有信奉宗教的作者才将神性立法者这种观点作为促成现代科学出现的因素。在解释为什么中国文明尽管有那么多崇高和杰出的成就,却出不了一个牛顿或笛卡尔时,坚决不信教的生物学家艾德华·威尔森指出这样一个事实:

> (中国学者)已经放弃了存在一个具有人性和创造力的上帝这样的想法。在他们的世界里不存在理性的自然之主,因此他们细心描绘的事物不遵循宇宙的原理……因为缺乏一种对于一般规律或者说"上帝的想法"的迫切需求,他们极少或根本没有去探寻这些规律或"想法"。

宇宙学家安德烈·林德自己是一名无神论者,但他也认为宇宙"所有的部分均由同一条法则支配"。这种观点与一神论有一定的相关性。

另一方面,仅凭一神论本身是不足以制造一次科学革命的。伊斯兰教是信奉一神论的,但是穆斯林世界的科学进步最终慢慢消亡了。虽然拜占庭文明信奉基督教，而且从未遗忘古希腊的科学，但自己却几乎没有形成科学。所以很可能是西方教会对法律、制度和理性特别强烈的重视，对现代科学的诞生起到了决定性的作用。

但是，在文艺复兴时期还有其他东西在起作用，那是一种创新的、孜孜以求的质疑精神，必须要看到事物本身，且对所接受到的观点持怀疑态度。对一些人来说，这甚至延伸到了宗教问题上，但是与很多人的想象相反，宗教怀疑论在科学革命中似乎并没有起到推动作用。几乎所有科学革命的伟大奠基人，包括哥白尼、开普勒、伽利略、玻意尔、胡克和牛顿都是虔诚的宗教徒。他们中的一些人，像开普勒和玻意尔很明显是受宗教信仰的启发而开始进行科学研究的，并没有一个是因为反对宗教信仰或正教才进行科学研究的。至少到 19 世纪中期为止，大多数伟大的科学家都是信教的，很多科学家至今仍是如此。

科学革命

科学方法

科学革命是以三大成就为特征的。第一是亚里士多德综合法的瓦解以及与其思辨性、推理性和定性的科学研究法的分离（虽然科学仍被叫做"自然哲学"）。第二是意识到实验和精确测量的重要性。这使得在自然调查中使用了更多的人造设备如望远镜、钟摆和真空泵。（这第二大成就也需要概念上的进步，在亚里士多德思想中，机器被认为会导致事物逆"自然"运动。）第三是科学的数学化。当然，之前的科学也用到数学甚至实验，但在17世纪这些工具才综合到一起形成了一种研究世界的有力的新方法，通常被称为"科学方法"。

科学方法包括：1. 通过精确测量和可重复的对照实验收集数据；2. 形成可检验的假设来解释数据中的规律性或反常性；3. 通过比较预期与新的测量结果或实验结果来证实或证伪这些假设。不幸的是，科学方法有时被说成是一种自动的或机械的过程。它不是一种"过程"，而是一种"行为"。过程是可以由机械来进行的，行为则需要个人的想象力、洞察力、智慧、主动性、创造力和判断力。

现在科学教育的主要目标是掌握大量的事实和理论，实验和观察是多么惊人的巧妙这一点却没有得到足够的重视。伟大的实验和伟大的理论同样美妙，但是它们似乎更稍纵即逝。它们是建筑科学大厦的脚手架，却经常在完成使命后就被遗忘了。

很多人对科学理论的形成有一种不切实际的理解。摆在学生面前的通常是作为成品的理论，所有含糊之处都被剔除，关键的想法被提炼为精确的概念，基本原理确定了，逻辑结构也理清了。这的确是学生掌握这些思想最好的方法，但是这也可能导致学生看不到最初得到这些想法所经历的混乱、困惑和痛苦的努力。

一个有力的事例可以证明：缺乏对真正科学方法的理解对哲学造成了有害的影响。欧洲17世纪和18世纪两次伟大的相互对立的运动——理性主义和经验主义，从某种程度上说都源于科学进步并认为自己在本质上是科学的。但是两者都夸大了科学思想的一极而忽略了另一极。理性主义者往往认为所有的知识都是像数学那样从最初的原理按演绎推理的过程发展而来的。他们低估了科学进步过程中经验成分的价值。另一方面，经验主义者没能正确评价假设、抽象和理论构建在促进科学发展方面所起的作用。和理性主义者一样，他们可能也被与数学的错误类比误导了。在数学中，每一个结论必须经过严格的证明才能作为下一步推理的基础。一些人以为经验科学的模式是类似的。他们认为理论假设存在的事物和每一个理论主张的事实都必须得到直接的证明才能进行下一步研究。但事实并非如此。

举例来说，电磁理论是詹姆士·克拉克·麦克斯韦在19世纪中期提出的。该理论假设在时空的每一个点上存在一个三分量"电场"和一个三分量"磁场"。没有人直接证明过这种存在于时空每一个角落的实体，也不可能这样做。但是，这是不必去证明的。那不是最初麦克斯韦理论有效性建立的基础或者物理学仍然认为其为真的原因。像所有复杂的科学理论一样，麦克斯韦理论有非常精细和抽象的结构，它预设了许多物质的存在，这些物质没有一个能被直接地或单独地观察到。相反，是这个理论作为一个整体可以通过观察被证明，但不管这些观察量有多大，与这个理论假设存在的实体相比必然是很少的。

经验主义者似乎认为，人类是通过将大量感官"印象"加总来构筑关于现实的图景的，而这些感官"印象"又通过一个"联想"的过程形成更复杂的思想。但是人不能直接感觉到磁场（除非假设说人是一只大王蝶），而经常是通过非常间接的现象，甚至在当时仅仅依靠抽象的理论推断它们的存在。但是这些磁场是真实存在的，就像岩石和树木那样实实在在。（电磁光谱同样说明了这一点：我们能用眼睛直接感觉到"可见光"，却只能通过间接的手段推断紫外线和无线电波的存在。但这种差别仅仅是因我们感觉器官的特征造成的，无线电波、紫外线和可见光本身都同样真实，不同的只是它们的波长。）我们可以在18世纪英国经验主义者身上隐约看到对于证明的粗糙的观点，这种观点也困扰了后来的各种经验主义，如20世纪早期叫做"逻辑实证主义"的哲学派别。对于逻辑实证主义者来说，每一种有意义的陈述都必须能够转化为感官印象的陈述。

在科学上，理论和观察的关系是复杂而动态的。理论建立在实验的基础上，而实验依赖理论对其进行解释。最近，关于"实验的理论依赖性"的讨论很多，这为一些"后现代"思想家声称科学方法导致了一种"恶性循环"，以某种方式破坏了对于科学客观性的看法提供了机会。通过一个简单的类比，我们就可以看穿这样的诡辩：地图是由探险者制作的，探险者又必须利用现有的地图。这种"循环"显然没有阻碍更好的地图的制作，理论和实验之间动态的相互作用也没有阻碍关于物理世界更好的理论的形成。事实上，这正是它们形成的必经过程。对于这一事实的认同是17世纪科学革命的根本成就。

在我们开始讲那次革命的历史之前，有必要再讲一下科学理论是怎样被证明的。在数学中，一个定理可以得到精确的证明，就是说它可以被毫无疑问地证明成立；或者它可以被证明不成立，即它可以被毫无疑问地否定。但是，就如同在生活中一样，在自然科学中，我们带着一定程度的信心接受一种理论，这种信心是，而且一直是一个不断变化的量。在一些情况下，这种信心非常强，我们可以说它是确凿无疑的，实际上是一个"科学事实"。例如，没有科学家还在认真地怀疑物质是由原子构成的或者地球在围绕地轴转动。在另一些情况下，科学家对一种理论的信心相当强，但还没有强到他们会说它已经被"证实"的程度。在还有一些情况下，在支持一个理论的时候，仅存在律师所说的"可驳回的推定"。这全要依赖于证据的数量和类型。

对于一个理论来说，什么能算得上证据呢？这不仅仅是经实验验证的定量预言这么简单。一方面，一个具体的实验数据，或者甚至是很多这样的数据，可能有多种解释的方式。例如，爱因斯坦引力理论（所谓的广义相对论）的重大成功之一在于它精确预言了"水星近日点的进动"（水星在经过距太阳最近的点时发生缓慢的位移）。但是，近日点位移现象还可以有许多其他解释。有一种是假设存在一定量的太阳扁率（如太阳两极的一个扁率）。另一种解释是假设在水星和太阳之间存在充满了一定密度的物质。现在，我们对太阳及其环境有了充分的了解，因此这些可能的解释都不再成立。但是有可能找到其他替代的解释，如可以简单地在牛顿的负二次方引力定律中加一项来解释近日点位移。所以，那次位移的成功预言虽然非常重要，但并不是物理学家开始相信爱因斯坦理论的唯一原因。

很多考虑的因素都会影响科学家对于一个理论的说服力和可能性的判断。这些因素包括理论的简明性和经济性，它提供一幅更统一、更清晰的自然图景的能力，它的解释力，它的数学美，它的深层理论基础，它对新现象的预见力，以及它解决理论难题或者矛盾的能力。例如，爱因斯坦的引力理论解决了牛顿引力理论和狭义相对论的原理不一致的问题。它还解释了为什么"惯性质量"等于"引力质量"。它预言了光线因引力而弯曲的现象。它源于深层次的"等效性原理"。它基于一个很美的观点，即引力是时空弯曲的结果。换句话说，爱因斯坦的理论是由许多条聚拢的证据线和可靠的依据所支持的。

理论如何被接受的另一个很好的例子是由保罗·阿德里·莫里斯·狄拉克于1928年发明的方程，它用一种与狭义相对论和量子理论都一致的理论来描述电子。狄拉克发明这个方程最初是出于数学美的考虑。但是这个方程也解决了一个难题，即电子的磁矩是之前理论所认为的两倍。狄拉克方程预言了一个新现象：反粒子的存在。它还阐明了电子自旋的特性。人们最后用它做了很多精确的实验预言，后来都得到了证实。这个例子说明了一个事实：虽然很多因素会导致理论家接受某种假设或对其建立信心，但最后通常是大量精准的定性的预言使他们打定主意。（至少在科学中是这样的，因为在科学中可重复的对照实验是可能做到的。但是，在所有的领域要求同样的证明是不合理的。例如大量汇集的证据证明发生了物种进化，但人们很难预言特定的世系是怎样进化的。同样的，人们可以知道地震的原因，却不能准确地预报地震。）

从哥白尼到牛顿

从某种意义上说,我们几乎可以说艾萨克·牛顿爵士(1643—1727)造就了科学革命。亚历山大·蒲柏著名的两行诗说得颇有道理:

自然和自然法则隐藏于黑暗之中。
上帝说,"让牛顿去吧",于是天地一片光明。

牛顿是一座高耸的山峰。20世纪以前,他在物理界无人能及。我们可以认为牛顿之前的一切都在为他的伟大突破做准备,而他之后的一切(直到20世纪)都是在利用他取得的突破。

三条发展线促成了牛顿的成就:在天文学方面,开普勒发现了行星运动定律;在物理学方面,伽利略发现了落体定律;在数学方面,笛卡尔(1596—1650)发展了解析几何学并开创了坐标的使用。

在天文学上,通向牛顿的那条线始于哥白尼(1473—1543),他以其行星运动日心说引发了科学革命。伟大的丹麦天文学家第谷·布拉赫(1546—1601)进行了极其精确的天文学观察,延续了这条线的发展。约翰内斯·开普勒的行星运动三大定律的发现使这条线的发展到达顶峰(他的发现如果没有布拉赫的数据是不可能的)。伽利略(1564—1642)在这条发展线上并不重要,事实上,他坚决反对开普勒行星椭圆轨道的重要观点。相反,伽利略对于天文学的伟大贡献在于望远镜的使用,通过使用望远镜他有了一系列惊人的发现,例如金星的相位、木星的卫星和太阳黑子,这些发现帮助摧毁了亚里士多德主义和托勒密体系。但是，在天文学理论方面是哥白尼和开普勒，而不是伽利略起了重要的推动作用。另一方面，在物理学上，伽利略通过实验（运用斜坡和钟摆）和数学的有力结合发现了落体定律，做出了重要突破。

在开普勒行星定律和伽利略落体定律中，我们看到了将精确的数学定律运用到特定系统中或小范围现象中的例子。今天我们会称这些为"经验关系"或"现象学定律"。牛顿的天赋使他能看到这些关系背后有更具普遍性、更具深度的法则在操纵，即万有引力定律和三大运动定律。

与科学上的所有伟大进步一样，这些定律的问世引发了深层次的统一。第一个统一是地球现象和天体现象的统一。牛顿以前，普遍存在的根深蒂固的观点是天地属于完全不同的领域，被根本不同的原理所控制，甚至是由不同种类的物质构成的。水晶般的天空似乎是永恒的，不受世俗世界特有的各种变化（"产

尼古拉·哥白尼（1473—1543），生于波兰的小城托伦，在克拉科夫大学学习天文学。他的舅父是埃姆兰的大主教，他为哥白尼在弗伦堡的教堂谋了一个教士的职位从事行政管理工作。哥白尼在博洛尼亚大学学习民法和教会法规，在帕多瓦大学学习医学，之后于1503年在费拉拉大学获得法典博士学位。此后，他回到埃姆兰做他舅父的顾问并当起了教士。作为一名天文学家，他声名远播，1539年，威腾伯格大学的数学教授乔治·约阿希姆·雷蒂库斯拜访了哥白尼，劝其将日心天文学的思想出版。在去世前不久，哥白尼完成了他划时代的巨著《天体运行论》（*On the Revolutions of the Heavenly Spheres*），出版的书在他临终时拿到了他的面前。这是一部引发科学革命的书。

第谷·布拉赫（1546—1601），出生于丹麦的一个贵族家庭。1560年，当他在哥本哈根大学学习的时候，一次被预言的日食激起了他对天文学的兴趣。在研究当时的星表时，他发现它们中没有一个和另一个相同。在17岁时，他决定"需要进行一个长期的计划，在同一个地点历经数年来绘制天体图"，为了这个（还有炼金术），他奉献了他的一生。

1572年，第谷观察到了一颗"新星"（一颗超新星）。他能够说明这颗"新星"在大气层外的很远处，这与亚里士多德天体不变的原则相矛盾。这让丹麦国王大为震动，为他建了一座天文台，第谷将它命名为"天堡"（意为"天空的城堡"）。后来，第谷在附近又有了另一所地下天文台，叫做"星堡"（意为"行星的城堡"）。

第谷用肉眼进行了最精确的（或是可以做到的最精确的）天文学观察。他反对哥白尼的日心理论，因为他知道地球围绕太阳转动会导致从地球上看到的恒星在天空中的位置发生微小的位移（"恒星视差"），但他观察不到这一点。（恒星距地球太远了，视差效应直到1838年才被观察到。）因此，他提出了自己的地心模型。在他的晚年，开普勒成了他的助手，并继承了他"宫廷数学家"的职务。开普勒从第谷那里继承下来的精确数据是一笔巨大的财富，帮助他发现火星的轨道是一个椭圆而不是圆，并用公式表达了他的行星运动三大定律。

第谷是一个异乎常人的人物。在学生时代，他在一次决斗中失去了部分的鼻子，在他的余生中都戴着用金和银制成的假鼻。在他位于克努兹斯图普的祖传城堡中，他经常大宴宾客，还养了一个叫杰普的宫廷弄臣，和一个他认为能预见未来的侏儒。第谷还养了一头驯服的驼鹿，有一天晚宴时驼鹿喝了太多的啤酒，后来跌下楼梯，死得很不体面。

生和毁灭"）的影响。因此，天体的"自然运动"按正圆轨道进行被认为是理所当然的，因为这样的运动周而复始，永无止尽（就像我们所看到的，甚至伽利略都没能完全摆脱这种古老的想法）。但是，牛顿告诉我们的是，控制天、地两界的是同一种力。行星绕太阳转动的轨道，钟摆的摆动和砝码的坠落都遵循相同的引力方程和力学方程。事实上，牛顿指出大海的潮汐可以由月球和太阳造成的引力来解释。而有趣的是，开普勒在很早以前就提出过这个观点，但却被伽利略嘲笑为迷信。

数学发挥新作用

科学革命以及它所带来的现代科学不仅仅是以数学和实验的愉快结合为特征，还包括人们开始用一种不同的方式看待数学及数学在物理世界的应用。哥白尼和伽利略时期的一个普遍观点是数学在定量描述事物上是有用的，但与理解这些事物或是它们的起因不是特别相关。例如，几何学技巧可以用来精确预测天体在特定的时刻会出现在天空中的何处，就像现代的火车时刻表能用来预测火车会在何时出现在各个车站。但是，就像火车时刻表不会告诉你是什么使火车前进或者火车为什么前进，许多亚里士多德主义者的观点是数学并没有关照到现象的内在以及潜藏其中的物理原因：那是"自然哲学"的任务。（在伽利略时代，科学家被称为"哲学家"而天文学家被称为"数学家"是有特殊意义的。）

这是哥白尼的日心体系在伽利略之前没有掀起多少波澜的一个原因。普遍认为，它仅是一种可选的计算方法，虽然有一定的优势，但并不涉及地球确实在转动这样的说法。人们普遍认为，在哥白尼体系中地球的转动仅仅是一种"假设"，就像是几何学家为了证明他们的定理所作的图,没有现实意义。只要任何一种计算方案能正确地预测天体在空中出现的位置,或按照他们所说,"保存这些出现",除了简便,它与其他任何方案没有什么差别。

现在,事实上,伽利略通过望远镜的一些发现(尤其是金星的相位)说明托勒密体系甚至不再能"保存这些出现"了,但哥白尼的体系却可以。但是,要证明地球的确在运动,那还不够,因为除了哥白尼体系和托勒密体系,当时还存在第谷提出的另一种可选的体系。第谷的体系可以像哥白尼的体系一样完好地保存

约翰内斯·开普勒(1571—1630),出生于德国威尔。为了成为一名路德教牧师,他进入蒂宾根大学学习,之后听说哥白尼的思想而迷上了天文学。他曾在格拉茨教数学,但是被势头正劲的天主教反宗教革命势力驱逐了。他又在布拉格找了一份工作,与第谷共事,但最终因为同样的原因被迫离开了那里。(开普勒的路德教教友也令他苦恼。因为他对圣餐的看法,他们将他逐出教门。还有一次,他们认为他的母亲会巫术而控告她。)开普勒利用第谷的数据发现了行星运动三大定律。他能做出这些发现不仅是因为他的坚持和他的数学技巧,还因为他良好的物理直觉告诉他,太阳作为行星系统中最大的天体,一定以某种方式对其他天体有控制性的影响力。他的天文学思想深受毕达哥拉斯神秘主义和基督教神学的影响,这使他看到了三位一体和天体之间相互关系的相似之处。虽然哥白尼和第谷给了他钥匙,但是是开普勒打开了通向新科学的大门。在他的书《宇宙和谐论》(*The Harmonies of the World*)中,他公布了他的第三条定律,最后,开普勒欣喜地说道:"我感谢您,上帝,我们的造物主,您让我看到了您的创造中的美。"

所有的出现，但不用假设地球在运动。（因此，它受到当时天主教天文学家的支持。）实际上，第谷体系不过是从地球的角度来看的哥白尼体系（或者按照我们今天的说法，它是在假设地球为静止的参照系中看起来的哥白尼体系）。从纯数学角度来看，如果我们像伽利略时期的大多数人那样理解数学的角色，也就是说，如果我们把数学和物理起因分离开来，是无法在哥白尼体系和第谷体系之间做决定的，即使是今天也仍然不行。

所有那些都因牛顿而改变了，因为他的运动定律和引力定律提供了运动的数学描述和运动的物理起因之间的关键联系点。具体说就是，它将加速与力联系到了一起。

要理解这一点，考虑一下这个简单的例子或许会有所帮助。假设一个体重200磅的男子用一根长的弹力绳拉着一颗小球绕自己做圆周运动。从数学角度来说，我们完全可以认为小球是静止的而那个男子在绕小球做圆周运动。为什么第一种描述从物理学的角度来说更合理呢？（事实上，明显是如此）原因是，在第一种描述中，我们明白了起作用的力，也就是"动力"。圆周运动涉及加速，如果知道小球的速度以及它轨道的半径，我们可以算出它的加速度。我们还可以计算绳作用于小球的力，因为我们了解绳。（具体地说，我们可以测量绳被拉伸的长度，还有一条叫胡克定律的经验定律可以将绳拉伸的长度和它作用的力相关联。）我们在球围绕人转这个合理的体系中看到，绳子上的力刚好等于球的加速度乘以它的质量，符合牛顿著名的定律 $F=ma$。

但是，在第二种描述中，球被认为是静止的而人在围绕球转动，我们不能以一种从物理学角度讲合理的方式来解释力。因为人的体重太大了，绳子的力不足以使他做圆周运动。因此，为了在假设小球静止的参照系中满足牛顿定律，必须假设有很大的附加力作用在人身上。它们来自哪里呢？不知道。它们没有明确的物理来源；它们必须专门介绍。用现代术语来说，它们是"虚拟力"，当我们不在一个物理学上合理的（或"惯性的"）参照系中描述一种状况的时候就需要假定它们的存在。

那么注意，如果有正确的"动力定律"，我们也许可以开始通过数学分析理解物理现实和物理起因。只有通过测量绳子的拉伸度，它的"弹力系数"，小球的质量和速度和它的轨道半径，然后通过必要的数学计算手段将牛顿力学定律运用到这些量上，我们才能从物理学的角度对正在发生的现象及其原因有一个正确的理解。（同样的，对牛顿的引力定律和牛顿运动定律的了解让我们看到，的确是地球在围绕质量比地球大得多的太阳转。）

这次科学的数学化远比牛顿之前的人所理解的要深刻，除了一少部分人可能对它有模糊的想法，如开普勒、伽利略，或许还有哥白尼。当伽利略说"自然这本伟大的书是用数学语言写的"，他预言了物理科学研究的一种全新的方法。

牛顿物理学

牛顿的定律 $F=ma$，指的是一个单一的点状物体，质量为 m，加速度为 a，受力为 F。但是，它可以被用到"扩展对象"和"连续介质"上——如果把它们想象成是由许多小的（效果上为点状的）部分构成。所以，牛顿力学可以被用来分析大量不同种类的现象，包括液体的流动、气体的压力、热量的流动、声音的震动和弹性固体的压力和张力。在牛顿定律形成后的两个世纪中，这些应用成为现实，物理学实现了一次更大的统一。

牛顿定律的普遍适用性和巨大解释力，以及将事物分解为部分进行分析的技巧促进了"机械论"世界观的流行。正如我们所看到的，宇宙是一台机器这种观点在中世纪思想中就已经很常见了。但是，中世纪的思想家最初认为天体运动是类似时钟的运动。17 世纪开始占主导地位的观点是万物，包括植物、动物和人体本身从本质上来说都可以用机械的术语来理解。

这种机械论观点的一个重要方面是"决定论"。物理学中的这种决定论观点不是来自于任何人的哲学偏见或预设，它来自于牛顿力学的数学本身。在牛顿力学中，一个物理系统在某个特定瞬间的状态，完全可以由一套数字，包括"坐标"和"动量"来表示其特征。坐标指明系统各部分在那一瞬间的位置，动量表明它们的瞬时速率。有了某一时刻的所有这些信息（"初始参数"），我们可以根据所谓的运动方程计算这些坐标和动量将会如何随着时间变化。而且，总的来说，这种变化是唯一的。这和象棋不同，比方说，下象棋是从某一个特定的初始状态开始，象棋的规则允许有很多不同的下法。在牛顿力学中，如果我们知道了某一时刻的位形（换言之，所有的坐标和动量），从那一点开始（或倒回）余下的"棋局"就以唯一的一种方式被确定了。那就是为什么1819年伟大的数学家和物理学家皮埃尔·西蒙·拉普拉斯（1749—1827）会写道："如果有一种智慧生命能够知道使自然具有活力的所有的力和构成自然的所有物体在某一瞬间的状态，那么没有什么是不确定的，未来还有过去都会呈现在它眼前。"

艾萨克·牛顿爵士（1642—1727），出生于英格兰的埃尔斯索普，1661年进入剑桥大学学习。他的天分很快得到了一位卢卡斯数学教授艾萨克·巴罗的认可，他在1669年辞职让位于牛顿。在他毕业后的1665—1666年，因为剑桥爆发了一场瘟疫，牛顿在埃尔斯索普度过了18个月，这或许是他科学家生命中最硕果累累的一段时间。就是在那段时间，他想出了颜色理论，发现了微分学和积分学的原理，并有了最终帮他形成引力理论和力学理论的重要见解。

由于对反射望远镜的改进，牛顿在1671年被选举为皇家学会会员（英国最高的科学荣誉）。之后不久，他提出了白光是由多种颜色光的混合这一发现，导致了激烈的争论，这使他后来放慢了发表研究成果的脚步。然而，在天文学家埃德蒙·哈雷的鼓励下，牛顿在1684年发表了他的巨著《自然哲学的数学原理》(The Mathematical Principles of Natural Philosophy)。现在普遍认同牛顿和莱布尼茨分别各自而且几乎同时发现了微积分；但是，谁是微积分的第一个发现者这个问题很快在当时引起了一场激烈的争论，这场舌战在他们死后还持续了很久。1704年，牛顿的《光学》(Opticks)出版。

从摒弃在亚里士多德科学中扮演非常重要角色的"目的论"这个意义上来说,牛顿物理学也是"机械论的"。就是说,在牛顿物理学中,一个系统的行为可以被预测,而不用涉及任何"终极因"(任何它正趋向的未来的"结局"或它正为之努力的"终

皮埃尔—西蒙·拉普拉斯侯爵(1749—1827),出生于法国诺曼底的博蒙。数学家让·勒朗·达朗贝尔非常欣赏他的才能,在他18岁的时候就为其在巴黎军事学院争取到了数学教授一职。通过短期内发表大量关于数学和数学天文学难题的论文,拉普拉斯24岁时在科学院获得一席之地。他的巨著《天体力学》(*Mécanique céleste*)是项大工程,五卷书出版于1799—1825年间。在这些书中,他用自己和他人尤其是他的朋友、伟大的数学家达朗贝尔精湛的数学技巧,将牛顿定律运用到详细了解太阳系运动这项难度极大的任务中。他最伟大的成就之一是证明了太阳系的稳定性。(牛顿认为某些不稳定性偶尔需要上帝的调节。)当看到《天体力学》这本书时,拿破仑问为什么书中从未提及上帝,对此,拉普拉斯著名的回答是"我不需要那种假设。"(据说当达朗贝尔听到这个说法时感叹道:"啊!但它是一个多么美的假设呀,它解释了很多事。")在他的《概率分析理论》(*Théorie analytique des probabilités*)一书中,拉普拉斯还帮助奠定了概率论的基础。

在哲学方面,拉普拉斯是以他阐述清晰的物质决定论出名的。在政治上,他是一个游刃有余的人,总能赢得当权者的喜爱。这使得他能够在法国革命中保全性命,后来又在拿破仑手下当了6个星期的内政部长。拿破仑在他的回忆录中写道,拉普拉斯曾是"一个够不及平均水准的管理者,为了一点小事到处搜查,还把无限小的精神用到了政府事务中"。

点")。相反，知道系统的过去状态和物理定律就够了。这帮助形成了自然是"盲目的"且无"目的"的这个观点。但是，值得注意的是，也可以用一种稍微更加"目的论"一点的方式看待牛顿物理学。在18世纪和19世纪，主要通过皮埃尔·路易·莫佩尔蒂（1698—1795）、莱昂哈德·欧拉（1707—1783）、让·勒朗·达朗贝尔（1717—1783）、约瑟夫·路易斯·拉格朗日（1736—1813）和威廉·卢云·哈密顿（1805—1865）的作品，根据所谓的"最小作用量原理"重新阐述牛顿力学的方式得到了发展。在光学中，皮埃尔·费马（1601—1665）在一个世纪之前阐述了一个相似的原理，叫做"最小时间原理"，指的是一束光在从某一始点到达某一终点的过程中会按照耗时最短的路径通过。因此，要用这个原理来解决光的路径，必须事先知道光的始点和终点。力学中相似的原理是，任何系统都会按照一定顺序的中间位形（被称为"轨道"、"路径"或"历史"）从初始位形向最终位形发展，而这些中间位形使得叫做"作用量"（通常记做S）的一个量最小化。

从它给出的答案相同的意义上说，这种表达牛顿力学的方式和之前从力的角度阐述的方式在数学上是等价的。但是，根据作用量原理的阐述更美、更有力，也更深刻。根据之前的表述，我们从用来明确系统状态的坐标数量相同的很多个"运动方程"开始着手（对一个复杂系统来说，这个数字可以是极其庞大的）。但是，根据"最小作用量原理"我们从根本的一个量着手，然后按照轨道使作用量最小化的要求就可以列出所有的运动方程。这样，我们看到了另一种正在发生的统一：许多定律（或方程）来自于一个只涉及一个根本量的力学"原理"。

力和场

虽然地心引力本质上是到目前为止最弱的自然力,但它却是控制远方天文学事件甚至地球事件的力量。原因是引力总归是有吸引力的,作用于地球上某物(例如,你)的引力都是地球大量原子相加在一起。相反,电磁力可以是引力也可以是斥力,物质中的正负带电粒子的作用往往几乎正好相互抵消。因此,即使电磁力实际上在我们可以直接观察到的大多数现象中起了主要作用,但它们不会像引力那样让我们在日常生活中明显地感觉到它们的存在。因为电磁力更难以找到,而且在数学上比牛顿引力更加复杂,许多科学家经过很长时间的努力才解开它们的秘密。其中最重要的几位是夏尔·奥古斯丁·库伦(1736—1806)、爱德华·卡文迪什(1731—1810)、亚历山德罗·伏特(1745—1827)、安德烈—玛丽·安培(1775—1836)、汉斯·奥斯特(1775—1851)、格奥尔格·西蒙·欧姆(1789—1854)和迈克尔·法拉第(1791—1867)。詹姆士·克拉克·麦克斯韦可能是19世纪最伟大的物理学家,他对这些人的发现进行了整理、扩展,并将其发展为一个统一的合乎逻辑的理论。

在麦克斯韦理论的发展道路上,关键的一步是"场"的概念。(有趣的是,提出这个至关重要的理论概念的不是理论学家,而是法拉第,一个历史上伟大的实验主义者。)牛顿的引力理论是建立在"超距作用"这个观点之上的。就是说,一个物体经过没有任何媒介的空间直接将引力作用于另一个物体。相反，法拉第认为，所有空间中都布满了电力场和磁力场。电荷和电流产生了这些场，同时又受到这些场的作用。我们可以认为电场是由"力线"构成的，力线从正电荷延伸到负电荷并用一种类似弹力绳的方式把它们拉到一起。

安托万·拉瓦锡（1743—1794），"化学之父"，出生于巴黎的一个富裕家庭。在学习法律之后，他的兴趣转向了科学。虽然进行了许多重要的实验，他最大的贡献是为混乱的化学理论建立起秩序。当时的化学家在一本本末倒置的理论的指导下苦苦研究。根据这种理论，物质被认为是通过向空气释放一种叫"燃素"的物质进行燃烧，而不是通过与空气中的某种叫做氧气的物质结合而燃烧的。发现氧气的约瑟夫·普利斯特认为氧气是缺乏燃素的空气；发现氢气的亨利·卡文迪什认为氢气是加了燃素的水，而氧气是缺乏燃素的水。拉瓦锡提出了燃烧其实是一个氧化过程，而燃素完全是一个错误的想法。在当时，很多化学家仍然相信四种基本元素——空气、土、火和水，但拉瓦锡制作出了一张非常精确的33位元素表（其中只有3个后来证明是化合物）。他为此前一片混乱的化学术语建立了合理的秩序。(氧化锌之前被叫做"锌的花"；氧化铁被叫做"能收敛的火星番红花"；氧化铅在英国叫"红铅"，在法国叫"铅丹"；硫磺酸被叫成"硫酸的油"；等等。）拉瓦锡弄清了盐、酸、氧化物等的区别和联系，并发明了现代化学术语体系。在恐怖统治时期极权主义的疯狂中，拉瓦锡受到了子虚乌有的指控然后被斩首。拉格朗日说道："仅仅一瞬间，我们就砍下了他的头，但是再过一个世纪也未必能有如此的头脑出现。"

后来人们发现，这些场有自己的生命。它们包含能量，而且不仅作用于物质的带电粒子，也相互作用。事实上，这些场和物质的粒子一样真实。当麦克斯韦完成他的理论时，他发现该理论的方程暗示波可以在这些场中传播且传播速度等于光速。实际上，后来的实验说明光确实是由这样的电磁波构成。这样，麦克斯韦的理论实现了长期以来被认为是完全不同的三个领域的现象即电学、磁学和光学现象的统一。实际上，涉及的统一远比这大得多，因为电磁力是原子之间和原子内部相互作用的

詹姆士·克拉克·麦克斯韦（1831—1879），出生于爱丁堡。1854年，毕业于剑桥大学三一学院，在阿伯丁的马修学院和伦敦的国王学院担任教授。1871年，被任命为剑桥大学卡文迪什物理学教授。在阿伯丁的时候，他写了一篇长68页的关于土星环本质的获奖论文，说明土星环只有以分离的粒子构成才能保持稳定。皇室天文学家乔治·艾里爵士说这是他所见过的最非凡的数学应用之一。这使得麦克斯韦开始考虑气体中的分子运动并运用统计的方法来了解它们的运动。他在不依赖玻尔兹曼的情况下发现了表示气体分子速率的"麦克斯韦—玻尔兹曼分布"，他还为统计力学和热力学作出了其他重要贡献。受到法拉第思想的启发，他之后开始研究电学和磁学。他得出了一个关于"场"和"力线"的力学理论，用四个现在被称为"麦克斯韦方程"的方程式实现了数学的完整性，这是19世纪物理学最伟大的成就。麦克斯韦是一个非常虔诚的基督徒，个性温和，为人谦逊，同时他也被称为是"最友善和幽默的同伴"。在晚年，他忘我地照顾他生病的妻子，直到他罹患癌症无力再照顾。在他84岁的时候，癌症夺去了他的生命。

原因。这么一来，"凝聚体物理学"研究的物质的物理特性（例如热传导性、弹性、不透明性、粘滞性等等）和物质的化学性质都是基于粒子间相互的电磁作用。

虽然麦克斯韦的理论涉及新的力、现象和概念，它在本质上还是一种源于牛顿学说的理论。和牛顿力学一样，它建立在一组方程之上（事实上，和牛顿的方程一样，它们是"二阶微分方程"），这组方程决定性地控制着一套坐标和动量随着时间的变化。当然，坐标的概念必须要拓宽，不仅要包括粒子的位置（就像牛顿所考虑的），还要包括存在于空间中每一处的磁场强度。然而，麦克斯韦的理论是对牛顿的概念和原理的一种延伸而不是遗弃。

20世纪的物理学革命

相对论

"运动相对性"的观点在爱因斯坦之前很久就存在,事实上它可以追溯到牛顿物理学。它的观点是速度不是一个物体的属性,而是两个物体之间的一种关系。也就是说,问一个物体的速度是没有意义的,我们必须问它相对于另一个物体的速度是多少。(一个有用的类比是,问一条线所成的角是无意义的,只有两条线所成的角才有意义。)换言之,在牛顿物理学中,速度是一个"相对的"概念。这可能看起来与之前所讲的可以使人通过对力的分析决定究竟是球在绕着人转还是人绕着球转的牛顿运动定律相矛盾。但是,其实不存在冲突:对力的分析不是反映哪些物体在运动,而是反映哪些物体在加速;"加速度"在牛顿物理学中是一个绝对的概念。(讨论单一物体的加速度是合理的,因为加速度是一个物体在某一瞬间相对于它在另一瞬间的速度。)

如果一个人在火车上,火车忽然开始加速,即使他闭着眼睛也可以知道,因为他可以感觉到一种力或摇晃。但是,如果火车不是在加速,人就不能分辨火车是静止的还是在以恒定的速度非常平稳地滑行。实际上,问火车本身是否"真的"在动是没有意义的;我们只能问它是否相对站台或其他某物在运动。

同样的，如果有很多物体相对于彼此做匀速运动（没有一个作加速运动），按照牛顿物理学，问哪些物体真的在运动也是没有意义的。实际上，我们做的是从方便的角度出发选择一个物体，然后参照它来观测所有的运动。这就是我们之前所说的选择一个"参照系"。如果一个人坐在火车里，以火车为参照物来观测运动是比较方便的；如果一个人站在站台上，以站台为参照物来观测运动是比较方便的。但是，"相对性原理"告诉我们选择什么参照系从根本上来说是无关的。

说它"是无关的"是什么意思呢？就事物表面上是如何运动的这一点来说，它当然有关。对火车上的人来说，站台是运动的，但对站台上的人来说，火车是运动的。更专业地说，物体的坐标和动量在不同的参照系中是不同的。在牛顿物理学中有一条精确的规则，叫做"伽利略变换"，它确切地告诉我们它们的差别在哪里。[这条规则与我们的日常经验和直觉相符。例如，如果一辆汽车以每小时40英里的速度经过你，另一辆汽车朝着同一方向以每小时75英里的速度经过你，那么对于第一辆车里的人来说，第二辆车似乎是在以每小时35（75减40）英里的速度前进。这就是伽利略变换，似乎它就是常识。]

相反，"选择何种参照系是无关的"这种说法意味着不管我们使用何种参照系来测量坐标和动量，这些坐标和动量会遵守同样的方程。换言之，特定物体的运动在不同的参照系中看起来会不同，但物理法则在任何参照系都会有同样的数学形式。那是相对论原则的关键点和真正本质。

麦克斯韦的电磁理论似乎违背了相对性原理。似乎麦克斯韦方程只有坐标和动量在一个特殊的参照系中测得时才成立。那似乎给"绝对速度"提供了一个自然的定义，即在那个特殊参照系中测得的速度。因此，看起来不是神圣的相对性原理出错了就是麦克斯韦的电磁理论有问题。

就在这时，爱因斯坦在1905年进入了这个研究领域。他的目的其实是非常保守的：他既不想放弃相对性原理也不想放弃麦克斯韦理论。这迫使他迈出了大胆的一步。他认为看起来非常符合常识的伽利略变换是错误的，正确的变换法是由荷兰物理学家亨德里克·洛伦提出的。

阿尔伯特·爱因斯坦（1879—1955），出生于德国的乌尔姆。与传奇的神话不同，虽然爱因斯坦读书时在数学和物理学上表现突出，但是因为缺乏兴趣，他在需要更多记忆的科目上表现欠佳。1896年，他进入瑞士联邦理工学院学习并在1901年拿到学位。之后，未能留校任教的他开始在瑞士专利局工作。爱因斯坦对当时的物理学和它面临的主要理论问题有深刻的理解，他思考这些问题已经有好几年了。这在爱因斯坦的"奇迹年"最终开花结果，1905年，他发表了三篇划时代的论文：一篇提出了狭义相对论；一篇论述了叫"布朗运动"的现象（说明原子真的存在——这在当时还没有被普遍接受）；一篇论述了"光电效应"，在论文中，他通过证明光在某些情况下具有粒子行为而不是波的行为，从而帮助奠定了量子力学的基础。1909—1914年，他在苏黎世和布拉格担任教授。1914年他成了柏林大学的教授，并一直待在那里直到希特勒当权，他放弃了自己的德国国籍并接受了普林斯顿大学高等研究院的一个职位。1916年，他的引力理论——广义相对论出版。这是人类智慧的一座伟大丰碑。爱因斯坦知道他"站在巨人的肩膀上"（就像牛顿曾说的那样）。在他的书房中保存着三个人的画像：牛顿、法拉第和麦克斯韦。

如果不同参照系中的坐标和动量由洛伦兹变换联系起来，那么结果证明麦克斯韦方程在任何参照系中都成立。这样，相对性原理和麦克斯韦的理论就相互符合了。但是，有一个潜在的大问题：牛顿定律不再在所有参照系中都成立了！也就是说，爱因斯坦成功地解救了麦克斯韦，却牺牲了牛顿。结果是，牛顿定律必须要改动。

　　需要改动的牛顿定律究竟是什么呢？不是他著名的三大运动定律（包括 $F = ma$），不是能量守恒定律，也不是最小作用量原理。

　　所有这些在爱因斯坦物理学中依然成立。其实，只有一件事要改，那就是时空几何学。旧的伽利略变换是基于这样的观点：三维空间本质上是欧几里德几何体系的，而时间是完全不同于空间的。但是后来证明这些观点都是错误的。虽然在爱因斯坦之前没有人能理解它们真正的含义，但洛伦兹变换表明空间和时间一起构成了一个有着不同几何体系的四维流形。在这个新的几何体系中甚至毕达哥拉斯定理都要修改。

　　先考虑一下空间的三个维度是怎样相互联系起来的可以帮助我们了解时间是怎样和空间相联系的。我可以选择我的三个基本空间方向（或"坐标轴"）为"向前"、"向右"和"向上"。（这就是在空间中选择一个参照系。我可以用它测定某物的位置，例如说它是在我前面20英尺，右边30英尺，和上方10英尺。）但是，如果我将身体稍向左转，那么跟以前相比我就面向一个不同的方向了，我之前所谓的前方现在应该说成是偏前偏右方。类似的，在爱因斯坦的理论中，一个参照系中的"时间方向"在另一个参照系中就变成了部分的时间方向以及部分的空间方向。从这个深刻的意义上来说，爱因斯坦的理论统一了空间和时间。

相对性理论还导致了其他的统一。人们后来发现"能量"和"质量"从一定意义上来说是同一样东西（这就是著名的方程 $E = mc^2$ 的含义）。而且麦克斯韦理论中三分量电场和三分量磁场其实是一个单一的六分量"电磁场"的一部分，而不是独立的实体。实际上，在一个参照系中的纯电场在另一个参照系中是部分的电场和部分的磁场。

正如我们已经注意到的，牛顿的引力理论也需要修改。这也引入了一个关于时空结构的新假设，即四维时空的构造是曲线形的。正是这种弯曲或扭曲导致了爱因斯坦广义相对论中的所有引力效应。这使得引力更像电磁学，在电磁学中，引力不再像牛顿说的那样基于"超距作用"来理解，而是基于"场"来理解。（一个特定地点的"引力场"等于时空在该处的弯曲度。）这些引力场和麦克斯韦的电磁场一样有自己的生命，也可以让波在它们中间传播。所有的力都被认为来源于场的这个事实，为建立一个适用于所有力的"统一场论"创造了可能性。爱因斯坦在晚年试图找到这样一个理论但没有成功。但近几十年来，这个问题已经有了巨大的进步。

相对论有多大的"革命性"？

凭什么说爱因斯坦的狭义相对论和广义相对论是"革命性的"呢？当然，它们得出的结论是极度违反直觉的，同时也是非常惊人的。例如，它们说两件事"同时"发生不是绝对有意义的（即是相对的）：这依赖于参照系。但是，它们没有完全丢弃之前的物理学。事实上，就如我们所看到的，爱因斯坦得出狭义相对论的起因恰恰是他努力想要同时保留麦克斯韦的电磁理论和旧的相对性原理。因此，麦克斯韦的理论被毫无变动地保留了下来，这一点也不奇怪。而相对性原理本质上也没有变动。例如，和在牛顿物理学中一样，在爱因斯坦物理学中，速度是一个相对的概念而加速度是一个绝对的概念这一点是正确的。牛顿物理学中很多别的东西也被保留了下来，包括牛顿的三大运动定律，以及像力、速度、加速度、动量、质量和能量这些基本概念（虽然其中的一些量必须作为四维时空中的矢量，而不是三维空间的矢量来理解）。

把"革命"这个词用到科学理论上是具有误导性的。在一次革命中，旧的秩序被扫除。但是，在大多数物理学所谓的革命中，旧的观点不是简单地被丢弃，也不存在与过去的彻底决裂。一个比"革命"更好的形容科学中这些巨大进步的词是"突破"。在伟大的理论突破中，会形成深刻的、影响深远的新见解，这些见解会将科学的理解力提高到一个新的水平，一个更高的视角。但是，许多——实际上是大多数旧的认识依然有效，尽管在一些情况下他们要被新的认识调整或修正。也许物理学史上唯一一次真正的革命是第一次革命——17世纪的科学革命。那次革命之前的物理学，即亚里士多德的物理学，大都被束之高阁并被完全不同的东西代替了。

我们高中、大学和研究所的物理课程都不是从学习亚里士多德物理学开始，这个事实就可以证明这一点。亚里士多德物理学的细节仅是历史系学生，而不是作为科学工作者的现代科学家感兴趣的对象。现在进行科学研究完全没有必要知道亚里士多德的物理学。相反，在一个人开始学习相对论（或量子理论）之前，还是有必要花几年时间学习17世纪到19世纪的物理学的。那种物理学仍然起着很重要的作用。实际上，现代物理学和工程学的许多分支仍然仅使用相对论之前和量子论之前的概念。

而且，牛顿的力学和引力理论仍然是爱因斯坦物理学在速度相对于光速很小且引力场很弱的时候唯一正确的"极限"，这一点很重要。也就是说，当速度越小，引力场越弱，爱因斯坦的解和牛顿的解就越相近。一个旧理论是取代它的理论的正确"极限"这个观点非常重要。这样的话，旧理论严格来说是不正确的，但也不能简单地说它是错的。更好的说法是它"到某一点为止是正确的"。

有一个例子有助于更清楚地说明这一点。任何一张曼哈顿的地图，如果它是印在一张平的纸上，严格来说它一定是错误的，因为地球表面是弯曲的。但是，曼哈顿非常小，因此地球球度的影响可以忽略不计。（它对于一幅曼哈顿地图角度的影响不到万分之一度。）事实上，没有理由担心那些影响，因为它们与其他影响，如曼哈顿岛的多丘陵地形相比相形见绌。所以，如果讨论的是一个足够小的区域，忽略地球的球度是一种合理的近似。不能简单地说基于此的陈述是错误的；相反，它们包含了真实的信息而且正确地说明了地理关系。因此，对一种情形不完整、不确切的描述可能足以完整地传达关于那种情形的真实信息。否则，我们绝对学不到任何东西。

让我们举另外一个物理学"革命"的例子，这次革命还没有发生，但很多人预期它会到来。牛顿物理学和爱因斯坦物理学都建立在这样一个观点之上，即空间（或时空）是由彼此间相距为固定距离的点构成的一个不间断的流形。但是，量子力学使这些概念在极小物质上的适用性显得非常值得怀疑。许多物理学家预计我们凭直觉感知的时空概念在描述比一个叫做"普朗克长度"（约 10 厘米）的基本量级更小的东西时将完全不适用，全新的概念将会得到使用。也就是说，需要一个新的理论。他们认为如果发现了这种新的理论，它将表明我们凭直觉获得的"空间"、"时间"、"点"和"距离"的概念不再适用于物理世界，除非是从近似的意义上来说。当然，那种近似对于远大于普朗克长度的距离来说是非常好的，但是，不管怎样，它始终只是一种近似。假设有一天所有这些预期都被证明是正确的，那会意味着所有用到距离概念的陈述（如，"你手中拿的那本书长8英寸，宽5.2英寸，厚0.3英寸"或"从我家到办公室的距离是1.75英里"）都是错误的吗？当然不是。这种距离的概念，虽然严格来说只是近似有效，但在这种情况下，这是一种很好的近似，所以对它的使用吹毛求疵是迂腐的，也是不合理的。同样的，我们完全可以在很多情况下（包括所有那些出现在日常生活中的情况）继续理直气壮地使用牛顿物理学，尽管我们知道它不能对正在发生的事给我们一个确切的完整的解释。

　　最后一点需要强调的是：爱因斯坦的相对论与"一切都是相对的"这个愚蠢的观点毫无关系。在牛顿力学和爱因斯坦物理学中（就像通常在生活中），有些事是相对的而有些事是绝对的。例如，在爱因斯坦物理学和牛顿物理学中，速度都是相对的，而加速度都是绝对的。在牛顿物理学中，时间距离和空间距离都是绝对的，但在爱因斯坦物理学中，它们都是相对的，而一个叫做"时空距离"的东西是绝对的。而且在牛顿物理学中，真空中的光速是相对的，而在爱因斯坦物理学中它是绝对的（它在每一个参照系中都是相同的）。"相对性"这个术语已经造成了无尽的麻烦。事物在相对论中不比在牛顿物理学中更相对，而是相对的事物不同，绝对的事物也不同。

量子革命

量子理论不像相对论是一个人的智慧结晶，从 1900 年到 20 世纪 20 年代中期它的基本结构完成，许多伟大的科学家都为它的发展作出了贡献。量子理论主要的奠基人包括马克思·普朗克（1858—1947）、爱因斯坦（1879—1947）、路易·德布罗意（1892—1987）、尼尔斯·玻尔（1885—1962）、阿诺·索末菲（1868—1951）、马克思·玻恩（1882—1970）、维尔纳·海森堡（1901—1976）、埃尔温·施罗丁格（1887—1961）、沃尔夫冈·泡利（1900—1958）和保罗·狄拉克（1902—1984）。

"革命性的"这种说法用在量子理论上比用在相对论上要合适得多。相对论改变了我们对于空间和时间的理解，但是量子力学从根本上改变了所有物理学的概念框架。运用量子理论之前的概念框架的物理学理论（不管是牛顿的还是相对论的）被称为"经典"理论。

即便如此，简单地说量子理论"推翻"了经典物理学是一种误导。量子力学建立在经典物理学的基础之上，且受其影响很深。实际上，将任何经典理论"量子化"，即为它构建一个量子版本有一个精确的常规步骤。在近似范围内（大约说来，当系统很大时）量子版本给出的答案和经典版本给出的答案是一样的。而且，如果不用经典理论就不可能把量子理论的预言和实际测量联系起来。

关于量子理论的一个重要事实是它本质上是建立在概率的基础上的。概率在经典物理学中通常也很有用，但它们只是对实际局限性的一种调节。在经典物理学中，就如拉普拉斯所说，如果知道一个系统某一时刻的全部信息，我们就可以（从原则上来说）精确地知道它过去和未来的发展。所以不需要概率。但是，在量子理论中，一个系统的全部信息不能唯一决定它未来的动向——只能决定各种结果的概率。量子理论这种著名的不确定主义（或"不确定性"）显然在哲学上相当重要。有些人提出它与人类意志的解放有一定的关联，当然，这是一种非常有争议的说法。

量子理论的概率特征导致了非常困难的认识论和本体论的问题，这些问题已经引发了很多不同的解释。这些问题太复杂太微妙，无法在这里作评论。但是，有一点也许很重要，传统

维尔纳·海森堡（1901—1976），出生于德国的维茨堡。1923年从慕尼黑大学获得物理学博士学位后，他与格丁根大学的马克思·玻恩和哥本哈根大学的尼尔斯·玻尔一起致力于快速发展的量子物理学领域的研究。当时，基础物理学处在一片混乱之中，理论家试图努力找到一种适合量子观点的统一的体系来取代当时存在的将各种想法和方法东拼西凑的状况。1925年，在23岁的时候，海森堡发现了这个体系并出版了他的"矩阵力学"。（1926年，埃尔温·施罗丁格出版了"波动力学"，这很快被发现是在不同数学伪装下的相同理论。海森堡和施罗丁格都获得了诺贝尔物理学奖。）1927年，海森堡得出了他著名的非常重要的"测不准原理"，该原理认为经典物理学用来表示一个系统状态的"坐标"和"动量"不能同时取到确定值。他后来继续为核物理学、凝聚态物理学和粒子物理学作出了重要贡献。在第二次世界大战期间，他领导了德国的原子弹项目。他的动机和对该项目的义务一直是一个有争议的话题。但是毫无疑问的是，他谴责纳粹统治的"丑恶"，认为那是"假借政治运动的形式陷入了一种疯狂"。

的解释（也被称为"哥本哈根的"、"标准的"或"正统的"解释）特别重视"观察者"（即那个知道实验或观察结果的人）的思想。简单地说，这是因为概率和一个人的知识程度或知识缺乏程度有关。（如果一个人知道未来的结果，就不需要用概率来对它进行讨论。）正如著名的物理学家鲁道夫·佩尔斯爵士所说："量子力学的描述与知识有关，而知识须要依靠了解它的人。"佩尔斯和其他一些人，如诺贝尔奖获得者尤金·维格纳（1902—1995），曾提出量子理论传统的解释暗示了观察者的思想是不能完全由物理术语描述的。如果这是真的，那么这种说法有非常深刻的哲学含义，尤其是反唯物主义的含义。但是，对于传统解释的不满已经导致很多人转而支持其他替代的选择，如"多世界诠释"、"隐变量理论"或"导波理论"。就这些问题，不管是在物理学家还是在哲学家中，都没有被一个大多数人支持的观点。

所有这些哲学困惑并不意味着量子理论作为一种物理学理论有什么问题。对于它经得起考验的预言是没有任何歧义或争议的，在我写这本书之前的80年间，这些预言已经被不计其数的方法验证了。如果像许多前沿物理学家所预期的那样，超弦理论被证明是物理学的终极理论，量子理论很可能不受威胁，因为现在看来超弦理论似乎不涉及任何对于量子理论根本假设的修正。

我们已经注意到，物理学中最伟大的进步使我们对于自然的认识有了深刻的统一。量子理论也不例外。它引发了所有统一中最引人注目的统一之一，即物质和力的统一。在经典的麦克斯韦电磁理论中，光是在场中由波构成的。但是，普朗克在1900年及爱因斯坦在1905年提出，除非光被认为是以分散的块，或能量子，换言之粒子的形式传播，某些现象就无法解释。

这些光粒子现在被称为"光子"。量子理论解决了某些物质既可以是波也可以是粒子这种似乎自相矛盾的难题。之后,"波粒二象性"得到了全面的应用。经典理论认为是波的物质也被看成是粒子,经典理论认为是粒子的物质现在也被认为是波——实际上,场中的波就是如此。例如,电场中充满所有空间的电子既是一种粒子也是一种波。另一方面,就如法拉第告诉我们的,力也源自于场。这样,物质的粒子和它们相互作用的力都是同

迈克尔·法拉第(1791—1867),出生于伦敦。他是一个铁匠的儿子,用他自己的话说,他的教育"按最普通的描述,比普通走读学校的阅读、写作和算术基础知识好不了多少"。他在13岁的时候成为一家书店的役童,14岁开始7年书本装订工的学徒生涯,这给了他一个阅读许多科学书籍的机会。1813年,法拉第参加了著名化学家汉弗里·戴维爵士的一次公共演讲,做了大量的笔记。(戴维是用电将化合物分离的先驱,用这种方式他发现了5种化学元素。)之后,他申请一份能与戴维共事的工作,但遭到了拒绝。他很快又再次申请,寄给戴维他记的笔记。戴维非常欣赏他,雇他做秘书,但后来解雇了他(建议他回去做装订工),之后又再次雇他做实验室助理。很快,法拉第就凭自己的能力成了一名杰出的实验化学家。但他最为出名的是他的电磁学研究,尤其是证明线圈相对磁体移动会在线圈中产生感应电流。(他证明了不管是移动磁体还是移动线圈,产生的感应电流是一样的。这是使爱因斯坦发现相对论的一条线索。)法拉第的感应定律是麦克斯韦电磁理论的支柱之一。更重要的是,他第一个提出了力"场"的概念,这是现代物理学的重要观点。法拉第为人热诚、谦逊、慷慨,是伟大的科学家中最受欢迎的人物之一。

一种东西——场的表现形式，这就是为什么过去半个世纪基础物理学的基本语言被称为是量子场论的原因。两个粒子之间的力，可以像法拉第描述的那样被认为是因为两者之间延伸的场线的作用，也可以像理查德·费曼（1918—1988）描述的那样被认为是由于两者之间"虚拟粒子"的交换，这两种理解是等效的。

对称性的作用

对称性观点在现代物理学中扮演了重要角色。在数学和物理学中,"对称"这个词有一个精确的定义:如果某个物体的一种变形使它看起来和原先一样,那么那种变形就被称为是那个物体的一种"对称形"。比方说,将一片雪花旋转60°,它是没有变化的,那么我们就说旋转60°是雪花的一种对称形。雪花共有6种这样的"旋转对称",因为旋转60°、120°、180°、240°、300°或360°,它都不变。再举另一个例子,足球的图案有60种旋转对称,即足球有60种旋转方式可以使它看起来一样。数学一个高度发展的分支——群论就致力于对称性的研究。

在自然界最美丽的物质,包括花、贝壳和水晶中都能找到对称现象。它在艺术、建筑、音乐、舞蹈和诗歌中也普遍存在。我们可以在很多地方找到对称图形,如教堂的圆花窗、廊柱、横楣、瓦片图案、阿拉伯花饰、法国花园、舞步、舞者自己的编排及诗歌的节奏和韵律模式中,这些只是众多可能例子中的一小部分。对称有一种带给人美感的力量,因为它有助于一件事物的和谐、平衡和均衡,也因为它符合统一原则。为了实现对称,一个图形的所有部分必须有序地呈现。拿掉花的一片花瓣,雪花的一个角,列柱中的一根,对称就破坏了,统一也瓦解了。就像我们应该看到的,对称也是物理学中的一个统一原则。我们能在自然法则中看到更大的统一,部分是因为已经被理论家揭示的更深层更惊人的对称原则。

　　对称不是实体物质特有的，即使是像方程这么抽象的东西也可以有对称。事实上，物理学家最感兴趣的正是物理定律本身所具有的对称性。物理学的许多伟大进步已经使我们发现了这些定律新的根本对称。例如，狭义相对论的全部内容就是物理定律通过洛伦兹变换，在从一个"参照系"变换到另一个参照系时保持完全相同的数学形式。这样，爱因斯坦就等于是在说自然定律是"洛伦兹对称的"。另一个例子是，麦克斯韦的电磁学方程有一种叫"规范对称性"的微妙的对称，这是伟大的数学家和数学物理学家赫尔曼·外尔（1885—1955）发现的。实际上，值得注意的是，从本质上说电磁力的存在正是因为物理定律的规范性对称。

　　除了引力和电磁力外，20世纪还发现了另外两个重要的力，它们被称为"弱力"（或"弱相互作用力"）和"强力"（或"强相互作用力"）。因为它们只有在距离小于一个原子时才显著，我们在日常生活中碰不到它们。在20世纪60年代和70年代早期，人们认识到这些亚原子力也是以"规范"型的对称为基础的，而且实际上它们的存在正是因为自然法则中存在这些对称。规范对称在数学上是非常微妙的，和我们可以看到的那类对称，如雪花和花的对称大不相同。举例来说，强力下的对称从数学上来讲和发生在一个抽象的三维"复"空间中的旋转有关。（之所以称它为"复"是因为，在这样一个空间中用来定位某物的三个坐标不是普通的数，而是"复数"。复数是一个形式为 $a+ib$ 的数，其中 i 为 $\sqrt{-1}$ 。）

　　在20世纪70年代初期，人们进一步认识到三种非引力的力（即：电磁力、弱力和强力）可以被认为是一种"大统一"

力的一部分。（虽然这种假设还没有明确地得到证实，但是存在很多支持它的间接证据。）这些大统一理论是基于更引人注目的规范对称。例如，这类最简单的理论中的对称涉及抽象的五维复空间的旋转。除了规范对称，在基础物理学中其他极为深奥的对称也被认为非常重要。其中有一个叫做"超对称"，在它的数学公式中用到了所谓的格拉斯曼数。这些数有一个奇怪的特性，如果 A 和 B 是任意两个格拉斯曼数，那么 $A \times B = -B \times A$，而不是像普通的数那样 $A \times B = B \times A$。

正如一位知名粒子物理学家所写的：

> 对称性在我们对于物理世界的理解方面已经起到了越来越重要的作用。继旋转对称之后，物理学家构想出了更深奥的对称……最终设计中充满了对称这种信念支撑着基础物理学。
>
> 如果没有对称性的指导，当代物理学就不可能存在……物理学家向爱因斯坦学习，提出了对称性并看到可能存在物理世界的一个统一概念。他们听到各种对称在他们耳边低语。当物理学家远离日常经验而向终极设计者的思想靠拢时，我们的思想被训练得不再停留在它们熟悉的地方。

我们要了解的一点是：当代物理学理论，如大统一理论或超弦理论的数学结构非常繁杂，所以物理学家必须充分运用对称性来构建它们。它们不可能忽然之间被凭空想象出来，也不可能大费周章地通过一个接着一个地验证实验结果来构建。这些理论是由对称性指导的。

"数学无理由的有效性"

随着科学的进步，我们发现自然法则形成了一个统一的结构，实际上是一座集精妙、和谐和美于一身的壮观的大厦。我们对这种结构的了解越深入，就越发现它的深奥以及用巧妙的数学来描述它的必要性。值得注意的是，在被运用到自然界之前，那类数学大部分是由纯数学家纯粹因为其内在的趣味性和美而研究并发展起来的。例如，虽然和科学没有明显的联系，在19世纪初复数理论就已经高度发达了。但是后来在20世纪20年代证明，复数对于量子理论的构成是必不可少的。类似的，群论是在19世纪末20世纪初发展起来的，几年之后人们才发现它对基础物理学有用，实际上是至关重要。其他引人注目的例子可以举出很多。

这段历史使尤金·魏格纳开始考虑"数学无理由的有效性"（在了解物理世界时）。这里有一些更深层的秘密，似乎纯数学领域不是人类随意构建或是创造的，而是人类做出发现的地方。他们所发现的最美的观点中有很多已经被证明是物理现实最基本层次的典型或是样本。我们越深入研究自然的本质，就越发现它是数学化的。

从1984年开始，基础物理学家就为一种叫做"超弦理论"的理论所着迷，这种理论具有前所未有的数学深度。他们中有很多人认为这就是一直在寻找的所有物理现象的统一理论。在这种理论中，物质的构成不是粒子而是在十维时空中震动的弦圈。每一种粒子仿佛就是在这根弦上演奏的不同的音符。当代最伟大的物理学家之一约翰·霍根在向一名科学期刊撰稿人描述这个理论时，因为无法传达他的研究展示带给他的壮观而感到沮丧。他说："我觉得我没有成功地告诉你它的神奇，它惊人的一致性，它非凡的简洁性和它的美。"

世界深刻的数学和谐性是毕达哥拉斯在研究乐弦振动时首先发现的。2500年之后物理学的发展又回到了振动的乐弦，这是多么神奇啊！没有人能预想到毕达哥拉斯的理论被证明是正确的。

最后，让我们来想想赫尔曼·外尔1931年在耶鲁大学演讲时对于科学的数学简洁性和美所作的思考：

> 很多人认为现代科学和上帝无关。相反的，我发现今天一个有见识的人要从历史，从世界的精神面，从道德来接近上帝变得更难了，因为在那些方面我们遭遇了这个世界的痛苦和邪恶，这些很难和仁慈万能的上帝产生和谐一致。在这些方面，我们显然还没有成功揭开人性用来遮盖事物本质的面纱。但是在对物理性质的了解方面，我们已经很深入了，所以我们能看到一种与崇高理性相一致的无瑕的和谐。

心理学入门指南

引 言

　　心理学虽然被威廉·詹姆斯称为"恼人的小学科",但心理学却完整地涵盖了人的行为和活动,这些行为和活动似乎依赖和取决于,或部分依赖和取决于人的感知、思想、情感、动机和欲望。然而,这些心理过程又似乎依赖和取决于,或至少部分依赖和取决于内在生物状态和外在社会影响的双重作用。更为复杂的是,影响只有在被认为有社会效应的情况下才称之为"社会的";同样,只有在它们集中作用于人的动机、情感和欲望时才称之为"影响"。相反,这些心理过程也会反映并一定程度地受制于更大的文化、历史影响。总之,被詹姆斯讽刺为"恼人的小学科"的心理学事实上是一系列错综复杂、包罗万象的问题和观点。这些问题和观点都是缘于人类想要了解自我这个恒久课题。

　　当然,这个课题不属于任何特定学科,也不属于某群学者或科学家所特有。自我认知同时包括了生物、遗传、解剖学、医学、社会、公民、政治、道德以及美学等各种因素。这一系列客观事实和活动塑造、指引并定义了一个特定的生命。一个已结束和另一个刚开始的心理领域之间并不存在绝对清晰的界限。举个典型的例子,在某个领域的专家几乎确定的观点,另一个领域的专家却将其视为主要矛盾或问题。因此,政治科学家接受亚里士多德的名言"人是社会动物",并进一步着手研究政治团体的各种形式和建立基础。进化生物学家会观测社会生活所赋予人的适应优势；而心理分析学家却对人脱离社会环境的变化感兴趣。

毫无疑问，科学史上这些研究模式的专门化产生了大量的有用事实，以及理解广泛现象的基本原则。同样，我们应该清楚地认识到，这些成果是付出了代价的。狭隘的视角以及学术研究中"只见树木"、"管中窥豹"的倾向，是付出的主要代价。生物学家精确地总结出人体的原子分子组成，却没有概括出有关这个人体所属人的任何有兴趣的东西。神经生理学家对视觉神经中的信息加工过程做出了详细叙述，却没有解释为何某些场景令人激动不已，而另一些却平淡无奇。更不用说，负责的生物学家和神经生理学家只对那些经过缜密的试验所论证或是看似可行的推论做出论断。但是，那些毫不怀疑的信息消费者，尤其是在专家们夸大其词的论断所鼓励下，会极其容易相信那些"不折不扣的"谬论：视觉神经的传导是正常视力必不可少的，因此视觉就"只是"这些神经的传导！

以下将介绍心理学和相关学科专业领域中所获得的研究成果和理论。本书时常提醒读者，这些研究成果和理论是更大、更完整的故事中的一部分，而非整个故事。此外，只有更完整的故事才能让人最后判定故事每一环节的重要性。设想一下，例如《李尔王》或《战争与和平》的故事，当将它们作为一个整体来考虑时，哪怕李尔王的身份是荷兰国王而非英国国王，抑或皮埃尔没有小说中提到的那么高，很显然，这些小说的精髓不受任何影响。这一观点可以更简洁地表述为：除非人们对人类本性形成了全面且正当有理的认识，否则任何试图"窥一斑而知全豹"的行为都是幼稚的。

在此前提下，我们还必须清楚，任何事物的"本质"必须包括它的组成、功能及其展现自我的途径。比如，当你没有丝毫依据地将某物归入"人类"，或者声称人类的本性是"思想"，但却无法清晰辨别作为"思想"实例的行动或事件，此时却去讨论人性是很奇怪的。故事的片段虽不足，但只要它们不断迈向更完整的故事，它们仍发挥着至关重要的作用。

哲学心理学:学科的创立

古希腊时期(起源)

Gnothe se auton(认识你自己)——这是镌刻在德尔菲的阿波罗神殿上的一句格言。人们从这个古希腊世界中探讨心理学的本源,这一学科为人们献上了知识的希望。当思考大部分科学学术界中被关注的主题时,人们同样诉诸这个世界。为何如此?理由不计其数,但没有一个是完全正确的。将古希腊文化的繁荣,归功于其奴隶制经济允许富裕阶层从事学问研究,这种论点被证明是毫无依据的。首先,古希腊的奴隶制经济并未取得类似成就;再者,古希腊最伟大的学术成就并非来自那些贵族。事实上,小亚细亚的希腊殖民者才可以被称作是哲学思想的创立人,他们原居住于希腊大陆较富足的土地上。

对此,古希腊宗教可以给予一个更有希望且合理的解释,尽管难免不完全。希腊人生活在一个虔诚的世界,但他们并无国教;无正统教义的束缚,也无拥有"真理"的神父或上帝。奥林巴斯山的众神享有的只是拥有强大体能的不朽之躯。他们当中没有神是无所不知的,也很少有神会对凡人的琐事感兴趣,即使有,也是在极偶然的情况下。当涉及心理学、哲学和自然科学这些难题时,这些古希腊众神们即便不是一无所知,也会保持沉默。

据说，毕达哥拉斯（约前580—约前500）第一个自称为"哲学家"（*philosopher*）——一个爱好智慧的人，即由"爱"（*philos*）和"智慧"（*sophos*）组合而成。当然，对于提出的问题至今仍引起热烈的哲学讨论，毕达哥拉斯并非第一人。同样，古希腊也并非是第一个探讨人生、道德和社会问题的时期和地方。有充分证据显示，对这些问题的思考早已存在于更古老的文化——美索不达米亚文化、东方文化、埃及文化和希伯来文化。因此，面对古希腊文化，我们不应致力于寻找某类问题初现的时间。我们应当探寻在哪个时期，人们对这些问题形成了持续的、批判性的、怀疑性的分析模式，而正是这些分析模式继续不断地将哲学区分为一个独立的学科。

这种分析模式的核心是，要摒弃对传统、神示以及个人权威的信仰。以哲学方式来论述问题，意味着我们必须清楚，任何揭示的真理，或仅仅一个民族的习俗，都不可能成为任何基础议题的最终定论，更不用说社会上某人声称的智慧或灵感。在这样的理解下，哲学便可成为其他与其同样严谨的分析模

毕达哥拉斯（约前580—约前500），最早自称为"哲学家"。然而他却以令人炫目的数学才能和乐律发现而闻名。他的教学对柏拉图思想中一些主要原则的形成有着重要的意义。然而很少有人了解毕达哥拉斯的生活；他自己，就像苏格拉底一样，没留下任何著作。看上去他似乎是从萨摩斯岛而来，之后创建了克罗顿（意大利南部）的一个学术社团，同时也是一个宗教团体，一个科学学校，这里拥有众多忠诚的追随者。他最终被迫从克罗顿逃离到了马泰邦省，在那里度过了晚年。

式——其他学科——的发展基础，尤其是自然科学。

在心理学作为一门学科诞生之前，人们试图在预言家、圣贤和诗人那里寻求指引，以深入了解人类现状。古希腊时期，最频繁地被请教的是盲吟游诗人荷马（约生活在公元前700年）。荷马著有叙事史诗《伊里亚特》和《奥赛罗》，记载了人类几乎所有的情感和力量，以及取得胜利或导致惨败的不同战略方式。这些作品中出现了具有一定框架的"心理学"：生命以一种"灵"（*pneuma*）和"魂"（*psyche*）的形式出现，伴随着人心脏、胸部和咽喉的运动所引发的情感和动机。有一种"魂"可能会离开，但也会回归人体，载着梦的内容；有一种"魂"一旦丢失了，死亡也就随之而至。正如每个时代和文化中伟大的诗人或剧作家一样，荷马详述了当时人们的生活状况，而这是无法通过哲学或科学研究获得的。然而，荷马史诗始终是部充满想象力的诗作，正因如此，它似乎每回答一个问题，却也导致了更多的问题。

希波克拉底（约前460—约前377），一所影响巨大的位于希腊寇斯岛上的医学院的创办人。基于对病患的直接研究，他总结得出：感觉、运动和认知功能是由大脑控制的（而不是大家普遍认为的由心来控制）。他反对对疾病的非物理性解释，是第一位能够正确描述癫痫和肺炎等疾病症状的医生，而且其治疗方法基本上依赖于能促进自然痊愈的因素：自然空气、健康饮食、运动等等。希波克拉底的追随者写的著作超过60本，广泛涉及各类医学问题。他所开办的医学院要求学生进行道德宣誓，现在我们称之为"希波克拉底誓言"。

苏格拉底和柏拉图

就这样，早期的古希腊哲学家们广泛思索探讨着现实、宇宙以及物质本身的性质。苏格拉底（前470—前299）将哲学探讨的重心，由无生命的自然界本质转移到人类本性。也正是苏格拉底给心理学问题下了第一个清晰的定义，并首先给出了解决问题的暂行方法。这些问题广义上可分为：认识问题、行为问题和治理问题。对第一个认识问题的思考有：人们如何认识事物？真理有什么特征或迹象？获取真理的途径是什么？人们如何确定自己不受蒙骗或不耽于幻想？当两个关于"已知"的结论相互矛盾产生冲突时，该如何解决分歧？知识与信仰有何不同？知识与见解有何不同？

哲学学科中，这类问题形成了一个专门的研究领域，称作"认识论"（*epistemology*），即"知识"（*episteme*）的"逻各斯"（*logos*）。由于这一类问题都是针对人的感觉、认知、学习、记忆、判断和信念等基本认识过程所提出的，因而被认为是具有心理学意义的。苏格拉底似乎没有任何著作；他经常与不同的朋友或哲学家辩论，检验他人论点，在问答中不断提出挑战以揭露对方观点的错误和矛盾。这些对话都记载于最基础的哲学作品即柏拉图的《对话录》中。在这些作品中，苏格拉底和柏拉图（前427—前347）的智慧得以融合，后者重新编写并创作了整部知识探索的剧本，内容更为丰富、深刻。尽管认识问题没有一个直接且稳固的"解决方法"，但苏格拉底对这些问题的思考不断深入，将其浓缩为几个核心原则：

A. 这个可感知的世界瞬息万变、杂乱无章，而它的背后是个不变的领域。正是在这个领域，我们寻找真理。这点可以从数学中得到证明。任何拥有三条边和一个直角的实体或图形，才是直线三角形的真实形态，这是毕达哥拉斯的理论揭示出来的三角形的真实形态。现实中可感知的三角形各种各样，它的真实形态却是固定的、永恒的。

B. 如 A 中所讲，通过感官获取的知识必然是暂定的、含糊的、易变的。值得我们拥有的知识并不是在"自然界"中发现的，而是从人脑中自己的资源中提取而得。毕达哥拉斯发现毕达哥拉斯定理并不是通过对三边图像的观察而得，而是来自他对事物本质的深刻思考。因此，数学的真理储存于"心灵"本身，在哲学探索中被获取。

苏格拉底坚持的观点用现在的话可以理解为：抽象真理的领悟从根本上说是以理性认知过程为基础的，而不是那些与感知和学习相关的过程。这是因为，抽象的真理并不"存在于"某个特殊事物中，也不是任何经验的结果。因此，抽象真理的知识必然与经验无关，或者说，它是先天的。

行为问题可以改述为一个疑问：人们应该怎样生活？什么样的生活方式是正确的？这个疑问中嵌有诸多恼人的子问题。比如：为什么一个人要做到"至善"？什么是"至善"？除了获取快乐和避免痛苦之外，为何还有其他顾虑支配人的行为？人的行为究竟会不会受其他顾虑的支配？除了社会习俗和自身文化的价值观外，还有什么可以引导人们过正当的生活？某人正当的生活方式是否对其他人也适用，或者说不同的个体有不同的选择？在这些事情上又是什么人或事物来确定标准呢？当不同的文化价值观冲突时，人们如何判定孰对孰错呢？在价值观领域，有没有事物是绝对"正确"或"错误"的呢？或者，它是否都与文化本身有关呢？在哲学学科中，这些问题的思考产生形成了一个具体的领域，即伦理和道德哲学。同样，这些问题也为心理学研究的所有分支作了界定：作为行为根源的情感与动机；行为控制中奖与惩的作用；利他主义的有利条件；影响行为的社会因素；价值观的形成和及其在生活中的作用；人际影响的根源和本质；道德推理的本质及其影响因素。

正如柏拉图的《对话录》中所记载，苏格拉底在这些问题上的观点随年月有所变化。但其中保留的一些重要元素，仍会引起哲学家和心理学家关注：

A. 人类"灵魂"的本质是复杂而混成的：部分激情，部分意志，部分理智。情感驱使行动。行动也传达出人的意愿和

柏拉图（前 427—前 347），其《对话录》建立了哲学思考的明确领域：知识，道德和政治。他的学院可谓富足慷慨，因为它给他的学生亚里士多德提供了将近 20 年的学习条件。柏拉图身为雅典贵族，最初取名为亚里斯多克利（Aristocles），但是他的一副宽肩膀让他有了一个新绰号："柏拉顿"，意思为"宽肩膀的"。他的老师苏格拉底于公元前 399 年被处死，之后柏拉图走访了位于非洲和意大利的许多希腊的城市，在那里受到毕达哥拉斯思想的影响。公元前 387 年，他回到雅典并且建立了学院（这个机构在公元 529 年被东罗马帝国国王关闭）。在公元前 360 年，柏拉图游历至叙拉古，成为狄俄尼索斯二世的老师，后者是那个城邦的新国王，希望自己成为哲学家国王，并认为这样才是理想的国王。但是这番努力显然还是白费了。柏拉图历经万难回到了雅典，在那里安度晚年至 80 余岁，参加完一个学生的婚宴之后，在睡梦中去世。

欲望。行动也是对某些理性判断的回应。因此，"灵魂"总是受到敌对甚至是相冲突事物的困扰，就如一个奋力意欲到达目的地的车夫由完全不同的两匹马牵拉着：一匹俯首帖耳、循规蹈矩，另一匹却野性难驯、顽固执拗。

B. 正如在特定环境下的正当行为，正当生活中的统治权威就是理智，所有的情感和意愿均服从于理智。情感、意愿和理智三者之间的关系应该是和谐而有秩序的。这三者的不协调是一种疾病：一种只有哲学治疗才能治愈的精神失常。

C. 亦如知识本身，行为的首要理性原则——正当理性的基本规定——是无形的，不是来自"外在"，而内含于人类自身，需要用心地加以培养和修炼。当人们做到了这点，就是具有了美德。这样的培养应在人类童年时期就严加约束，让孩子置身于模范公民中，消除任何不良影响。当然，并非所有人都能达到期望的目的，因为它还需要某些源于基因的天生特质。

最后，《对话录》中提到的治理问题，为政治学专业领域及其子领域如法哲学、政治理论学或比较政治学奠定了基础。尽管哲学和政治科学已经深入探讨了这些领域，但它们几乎还未被心理学研究所涉及，这有必要形成心理学一个新分支：公民心理学（civic psychology）——研究在最基本的城市环境中生活的公民的心理层面。

柏拉图的《理想国》中大量探讨了苏格拉底对治理问题的思考，然而，这些对话并非始于探讨"善的城邦"（polis）的本质，而是"善的人类"（anthropos）的本质。由于后者需要被放大才能看得更清楚，因此苏格拉底认为，最好的方式是思考让城邦变善的方式。这也为"善的人类"本质的探讨提供模型。然而，值得注意的是，这部政治科学中奠基作品的开篇，就像是关于心理科学的探究。为了同伦理道德探究所取得结论保持一致，《理想国》辩称一个井然有序的城邦应由一位"哲学王"来统治，国王的法律应严格按理性原则管理民众，就如理想指导每位公民的行为。只要削减个人财产就会减少人的贪欲，甚至连小孩都应被大众共同"拥有"。同时，城邦的守护者应该以优生方式孕育而成。

亚里士多德及其自然观

很明显，对于那些会引起哲学、科学和心理学等领域进一步研究和争论的问题，柏拉图的《对话录》都涉及了。但是，大量的笔墨用于描写对话或辩论——甚至是辩论技巧——这使得柏拉图留给后人的成就很多，但却是混杂而古怪的。然而，古希腊还有其他成就，尤其是古希腊时期的医生，他们致力于解释和身体障碍相关的心理和行为异常背后的原因。希波克拉底（约前460—约前377）的学生和门徒们尤其值得称道。在他们对希波克拉底医术的解说中，他们建议观察脑疾病对人感知、思考以及行为的影响。他们反对"神圣"疾病（羊癫疯）的观念，并坚持认为每一种疾病应该从身体机能的层面去认识和理解。

然而，至今而言，对古代世界作出最为直接贡献的是亚里士多德（前384—前322）。他是物理学家的儿子，也是柏拉图学院认可的最优秀的学生。如果把柏拉图在《对话录》中展现的智慧比作是零散的剧作，那么亚里士多德的剧作则是学术的、系统的、循序渐进的、叙述详尽的——正是这些特征为一门学科划定了研究领域。就当前使用的"心理学家"这个名词的含义而言，亚里士多德是第一个心理学家，也是最伟大的心理学家之一；与其称他为哲学家式的心理学家，还不如称其为心理学家的心理学家。与纯粹基于思考的作品相区别的是，大部分亚里士多德的理论作品建立在直接观察的基础上。而他的心理学也更多地集中在对产生生物和环境影响的自然世界的观察上。在其主要作品中，柏拉图式的"灵魂"——一个脱离肉体的真理存储库，存在于人出生之前，也不因生命的死亡而消失——转化成为一系列真实存在的力量和过程，它们与身体及其潜能紧密相连。

正如亚里士多德所言，"灵魂"只是一个原则——生物(*zoon*)的基础或首要原则(*arche*)。在最简单的生命体中，灵魂被认为具有摄取营养和生殖的功能。生命的出生和存活，都必须有摄取营养的方法。物种本身的延续，则需要这个生命有生殖功能。就稍复杂层面而言，灵魂的功能还包括运动。植物向着太阳运

亚里士多德（前384—前322），物理学家之子，曾做过亚历山大大帝的老师，是古希腊最早的、最伟大的思想家之一。他的理论和研究至今仍是传统科学和人文学的重要基础。出生于色雷斯的斯塔基拉，17岁时被其监护人普罗克森努斯（他的父亲在他年幼时就去世了）送到雅典修习。他在那里进入柏拉图学院学习，并待了20年之久。在柏拉图去世后，亚里士多德离开雅典，最后到了马其顿国王腓力的王宫，成为年仅13岁的亚历山大的老师。腓力去世后，他回到雅典，开设了自己的学校——吕克昂；在那里他总是边走边授课，因而他的追随者被称之为"逍遥学派"。因为亚历山大的关系，他在雅典有不少优势。但是当亚历山大在公元前323年去世时，一场政变取代了雅典的亲马其顿政府。他不得不逃到优卑亚岛的卡尔西斯避难，之后不到1年就去世了。

动,正如同捕蝇草围捕和消化昆虫。就更复杂层面而言,除了营养、生殖和运动功能,灵魂还有感觉功能。亚里士多德认为,感觉功能是定义动物的关键。只要个体能够感觉,就被认为是有生命的。再进一步,亚里士多德提出了一种称作知性的力量,指基于经验的学习和记忆的能力。在动物界,可以发现这类功能的各种形式。然而,除了以上这些,成熟、健康的人根据自身能力还拥有一种特殊的力量:理性功能,这与知性有所不同。知性使生物可以学习并记忆细节;而理性,苏格拉底称之为"*epistemonikon*"(知识),使人领悟并形成普遍通用的观点。这就是真正的抽象思维的能力,它不仅是数学和逻辑学的基础,也是法制制度的根本。这种力量促使人们行动并理解行动,因而对抽象道德和司法戒律而言是有责任感的。当然,这种对理性功能的关注,并不意味着忽视对知识最初的感知。在诸多重要方面,亚里士多德都是一个注重实际的心理学家,他不会怀疑任何感官的功能,也不会忽视某个物种中广泛存在的能力或变化过程。

亚里士多德关于人性的理论通常被称为"形式质料论"(*hylomorphism*),由希腊词词根 *hule*(质料)和 *morphe*(形式)组合而成。亚里士多德认为,"灵魂"是"身体的形式"。而"形式"的具体所指,有必要看下亚里士多德的例子:如果说灵魂是一只眼睛,视觉就是它的形式。因此,雕塑上的手仅仅是"名义上的手",因为它并不实现"手"原本预期中的功能。因此我们可以理解为:生物在本质上是生命或行为的一种限定形式。从这个意义上说,人是一种理性动物;某些外观上极像人类却缺乏理性力量的实体只能是"名义上"的人。

亚里士多德创造出了一种系统的心理学,包括个体生物的、社会的以及政治的层面，并据此提出了一个全面的关于人性的理论。总理论体系中包含了一些子理论，有关于学习、记忆、动机、情感、认知和抽象思维的理论，还有关于遗传心理学、生物社会学、性别研究的理论。亚里士多德关于心理学的探究，其核心是一种广泛意义上目的论：自然世界的存在和特征，总是为着某一目的。大自然并非杂乱无章，而是存在着合理的联系。那么，对任何自然过程的理解，不能只局限于过程中起作用的物理原因，还应探究其"最终原因"，即关系网中它的构成及各种联系的最终目的。把人看是一种理性动物，其基本问题是人的理性力量是为何：它们的最终目的是什么。这里的理性，就如小鸟的翅膀。小鸟翅膀的存在是为了飞翔，那么，理性的存在是为了什么呢？亚里士多德认为，理性是为了维持生活的活跃状态，一种生活的幸福（*eudaimonia*）。这里的幸福并非指动物和孩子的感官快乐，而是指理性生活所拥有的深刻持久的幸福。它的完成程度，是衡量一个人德行或美德（*arête*）的标准，据此将一个人划入特定的类型。事实上，亚里士多德关于伦理的著述及其经典著作《修辞学》，为后来的人格理论奠定了基础。

几个世纪以来，所有给予心理学伟大定义的话题，几乎都为亚里士多德所言，而且亚里士多德更是其研究领域的开拓者之一，其理论为人们所引用。如果要给这个学科下一个简明的定义，那么只有亚里士多德的著作才可以为其定义提供基本术语：心理学是对感知、认知和（或）理性力量及其过程的研究，生物正是通过这些力量和过程与它们所在的自然和社会环境产生联系，以实现它们特有的本性所决定的目的。

亚里士多德之后的哲学心理学

从古希腊时期到17世纪，心理学发展主要得益于：(1) 哲学家采用内省法，研究他们自身思想和行动的本质，促进了心灵本质和心理世界的理论发展。(2) 医务人员记录了疾病和损伤对感知、思考、记忆和行动等心理学过程的影响。这段时期诞生了思想文化史上最具有创造性和最为敏捷的头脑，也见证了现代大学的创立。尤值得一提的是圣·奥古斯汀（354—430），他的《忏悔录》是一部真正关于心理过程的著作；圣托马斯·阿奎纳（1225—1274）是对亚里士多德最具有洞察力的评论家之一，他新颖、独创的思想涉及感知、认知和自由意志等领域。还有奥卡姆的威廉（1280—1349），他详尽叙述了哲学术语"共性"的认知基础。到12世纪，一些小型的教会学院得到扩大，从而可以提供各个学科教学，巴黎的学院开设了一套现在被称为"人文学科"的课程，开始享有了真正大学的地位。

17世纪中叶，心理科学作为一门独立的学科渐渐由可能成为现实。在这个发展过程中，三位最有影响力的人物是：勒内·笛卡尔（1596—1650），托马斯·霍布斯（1588—1679）和约翰·洛克（1632—1704）。他们被誉为科学的伟大时代即17世纪的代表，这也是开普勒、牛顿、玻意耳、伽利略、雷恩、惠更斯所在的时代。笛卡尔对数学和光学作出了杰出的贡献，被誉为"解析几何之父"。霍布斯在拜访了伽利略并掌握了力学的基本定律和原理之后，撰写了关于人性的奠基之作。洛克作为英国皇家学会的成员和一名医学博士，创作了哲学心理学史上最具影响力的著作之一：《人类理解论》（*An Essay Concerning Human Understanding*，1690）。很明显，洛克试图沿袭牛顿的自然科学来建立一门心理科学。根据洛克的论述，思想是一些基础感觉的合成，聚合在一起是因联想力的结果。在他看来，大脑完全"依赖于"经验，大脑的基本感觉与牛顿提出的光微粒相似，大脑的联想规律如同力学中的地心引力一样运作。

笛卡尔以其二元论而著名。他指出，思维是非物质的，与人体属性完全不同。笛卡尔关于知觉、情绪和动机的心理学理论完全是基于生理学层面的解释；人类的抽象理性力量则不在他的理论体系中。尽管在某种程度上，洛克的理论有些含糊，但他也明确倾向于认为，生理学是所有心理学过程的基础。霍布斯则是唯物主义心理学的无条件忠实拥护者。因此，在哲学心理学这个多产的时代，最具影响力的是那些反映自然科学发展的作品，同样，它们试图构建一门科学心理学。

笛卡儿（1596—1650），当时的奇才之一，在数学、光学和生物科学方面作出了巨大贡献。他的《方法论》(*Discourse on Method*)和《第一哲学沉思录》(*Meditations*)在哲学探究的本质和理性的首要性上建立起"笛卡尔"地位。他出生于法国土伦省莱耳市，就读于拉弗莱什的耶稣会学校。和当时大多数人一样，他相信理性和信仰是需要严格分离的。他在巴黎度过了年少时期，1629年，他归隐至荷兰长达20年，通过不断搬家来小心翼翼地保护私密。1649年，他收到瑞典克里斯汀娜女王的邀请并离开荷兰成为她的导师。由于不适应迥异的气候以及女王所要求的早起授课，他在到达瑞典不久后就去世了。尽管他的哲学著作中隐含对神学的质疑，但是他从未放弃天主教。他去世后名声大噪，很多人膜拜他为圣人；当他的遗体被运回法国时，由于追随者众多、热情，他的遗骸因不断被追随者收集而变得非常轻。

在下一个世纪，即启蒙运动时期的18世纪，对牛顿自然科学的信奉几乎达到了宗教门徒般的虔诚。物理学，尤其是应用物理学的发展，鼓励并坚定了人们的信念，即无论多么复杂的问题，都在自然科学的解释范围内。"社会工程"[1]出现在当时的伟大著作之中，有时候以宣传册和论文的形式问世，有时候以更为有力的革命辞令和动乱而引人注目。启蒙运动时期的政治，在很大程度上是一种政治心理学，它基于"正确"的人性论，认可某种治理模式同时排斥其他模式。对于洛克、大卫·休谟和英国其他经验主义者来说，事实和原则问题的解决并非通过传统或教义，而应通过经验。从笛卡尔的遗作中可以看出，启蒙运动时期的作家对认识世界的各种观点持怀疑态度，他们对于一切未经系统观察和理性分析的观点统统予以否定。但生命现象并非如此简单，社会和政治生活的各个领域都不断证明，启蒙运动时期的这种"信念"多半都站不住脚。

同样在这个世纪，医学尤其是神经病学的进步，为临床上心理和身体、大脑和思想关系的研究作出了较大贡献。18世纪末，颅相学之父弗朗茨·高尔（1758—1828）做了一系列具有说服力的解剖观察，从而得出结论：特定的心理过程取决于大脑皮层的特定区域。该观点提出后的几十年，人们致力于实验测试，以证实在哪里以及在什么程度上，特定的大脑皮层对应某种心理过程。就这样，人们开始投身于大脑功能定位的系统研究并直到现在。

---

[1]社会工程（social egngineering），暗指政府或者强势集团利用宣教、操控文化、法律制度企图改变或者"重整"平民——译者注。

作为自然科学的心理学

对于心理学什么时候离开哲学的停泊区而独自启航,并没有一个确定的时期。事实上,从这个学科的本质来说,它的"独立"更像个口号。自然科学,包括物理学的发展,都是基于一些假设和倾向,但这些假设和倾向并未被科学论证,也未被科学所蕴涵。看起来我们必须首先解决基本问题,才能开始知识探究。比如,物理学的定义是对物质、能量以及运动规律的研究,那么,必然需要某种基本理论来维护这个定义——某种可以确定一种关系为"规律"的基础,某种可以确定一种实体为"物质"的基础,某种可以确定一种势力为"能量"的基础,等等。总而言之,每个具体的科学研究领域都必然有其形而上学的基础。这里的"形而上学"是指本体论(对存在或本原的研究和思考)和认识论(对知识—论断以及获得知识方法的关键性检验)中相互关联的哲学问题。没有任何一门自然科学可以脱离这些基础而独立存在,否则就如同一扇没有铰链的门。然而,却有一大部分当代心理学家仍保持着对这个真理的无知或抵制。但是,也有另外一种意义上的"独立",这是由19世纪的一批科学家提出来的。他们急于想把他们的研究课题,从他们认为是无谓的哲学争论中脱离出来。其中对这一种"独立"最有影响力的辩护者包括来自德语世界的赫尔曼·冯·亥姆霍兹(1821—1894),英国的约翰·斯图亚特·穆勒(1806—1873)。他们分别用不同的方法制定了实验科学的原则,使之适合于思维过程的研究。亥姆霍兹将其理论建立在心理学的基础之上；穆勒则表示反对：他坚持认为研究"心理规律"的实验科学应该有别于研究"身体规律"的方法。不过，他们彼此都影响创造了一种氛围，促进了科学心理学和实验心理学的发展。

新科学的先锋包括莱比锡大学的威廉·冯特（1832—1920）以及哈佛大学的威廉·詹姆斯（1842—1910）。他们都在各自的大学中建立了实验室，用来研究基本的知觉过程和心理过程。尤其是冯特，他的莱比锡实验室建立于1878—1879年，尽管詹姆

约翰·斯图亚特·穆勒(1806—1873)，被公认为是19世纪英语国家中最主要的哲学家。在以经验主义替代理性主义的辩护上颇具权威，并且具有持久的影响力。在其逻辑著作中，他发展了实验科学的正式原则，并以此指导研究。作为哲学家詹姆斯·穆勒的大儿子，约翰·斯图亚特所接受的教养是极其严格的。比如，他在3岁的时候就学会了希腊语。青少年时期，他受杰里米·边沁的影响极大，并提出了"功利主义社会"。但父亲极为苛刻的教育方式使他精神分裂，他在21岁时就表现出严重的抑郁症。病愈后，他开始了自己的职业生涯。1823—1858，他在英国东印度公司工作，从一个普通职员升任为检查部门主管。1851年，他和一个名为夏洛特·泰勒的寡妇结婚，后者长久以来一直是他学术思想上的伴侣之一，有时候两人同著文章。1865—1868年，他服务于政府，任国会议员。尽管拒绝将自己的钱花在竞选之上（也就是通过金钱可以买到自己的席位），他还是赢得了议会席位。在去世（于法国城市阿维尼翁）之前，他勤勉写作，著有为数甚多、涉及甚广的文章，这为其赢得了"政治进步论者"的名声。

威廉·冯特(1832—1920)，现代"实验心理学之父"。他建立了世界上第一个心理实验室（莱比锡，1878—1879），创办了第一份发行广泛、刊登心理学研究的杂志。他的学生多被顶尖大学争相邀请去建立"德国"模式的心理学院系。作为马丁·路德教派神父的儿子，他出生于德国的一个叫做内卡劳的小村庄。他是一位训练有素的生理学家，曾经担任过赫尔曼·冯·亥姆霍兹的助手。1982年，他在海德堡大学第一次讲授心理学。他的大多数著作都是关于感觉的研究。但是，他也高度关注意识经验来辨别意识的"结构"和基本构成。结构主义是心灵主义派研究心理的方法，但这一理论后来为美国功能主义派和行为主义派所拒绝。

威廉·詹姆斯(1842—1910)，美国最伟大的哲学家，最敏锐的心理学家。他的《心理学原理》至今依然独领风骚。1875年，他在哈佛大学开设实验心理学课程，建立起美国第一个心理学实验室。1842年出生于纽约，和小说家亨利·詹姆斯是亲兄弟。他放弃了最初的绘画爱好，1864年进入哈佛大学就读医学，1869年硕士毕业，但那个时候遭遇了严重的抑郁期。1872年，他开始在哈佛教授本科生生理学课程，之后又开始教授心理学和哲学。1898年，他首次表示自己是实用主义者，该学派现在与他最常联系起来。但他的兴趣不止于此，因为对宗教和超自然持开放态度，他的研究在当时独树一帜，特立于心理学这个新兴领域。同时，这一兴趣也让他在1901年受到邀请，前往爱丁堡大学做吉福德讲座。翌年，这些讲座以书名《宗教经验之种种》(*Varieties of Religious Experience*)刊登出来，至今被奉为经典。1910年，他因心脏衰竭去世。

斯在他的书中提到他的哈佛实验室早在1875年就建立并投入使用。但若以其重要性而不仅仅依时间的先后来评价，冯特更值得称道。冯特创办了一份可以发表心理学研究成果的杂志，并且建立了研究生课题项目，对那些将在欧洲、大不列颠和美国任何大学里确立心理学项目的学生授予博士学位。在这些方面，冯特称得上是"实验心理学之父"，尽管威廉·詹姆斯的著作《心理学原理》(Principles of Psychology, 1890) 是该理论研究最为成熟的表述。这门新科学的特征随着心理学领域的各种发展而不断形成，其中重要的影响有两个：一个是特殊发展，另一个是积累发展。前者来自于查理斯·达尔文（1809—1882)的旷世之作；而后者的作品则是通过连续几十年的代代相传。了解它们是如何与后来出现的心理学学科相结合的，对我们具有指导意义。

达尔文进化论

达尔文的著作《物种起源》(Origin of Species, 1859)、《人类的由来》(Descent of Man, 1870) 和《人类和动物的情感表达》(Expression of the Emotions in Animals and Man, 1871)，从一开始出版就主导着心理学的思考。为了维护心理发展理论的连续性，并且根据这一理论将人类心理力量在动物界中真实地反映出来——尽管这种反映不尽完全或有所改变，达尔文还是不断促进对动物适应性行为的研究，并且增加其可信度。研究项目着重于对本能行为、交配和性选择、从幼小到成熟的发展过程、物种比较和异种等问题的研究——这是一套完整的理论——看上去像是一夜之间迅速发展起来。现在我们可以从物种选择的压力及其生存挑战的角度来理解社会制度和社会实践。人种比较则被认为是由于进化程度的不同，这使得种族主义者关于社会经济分层的解释变得可接受和可预测。心理疾病也用适应理论来理解。以往人们长期所持的"本质主义"(essentialism)认为，一种事物之所以是事物，是因为其亘古不变的本质，现在这个理论让位于"情境主义"(contextualism)，该理论认为万物必须适应生存状况才能得以生存。一个极其重要的观点更倾向于一种心理理论形式"功能主义"(functionalism)：心理学的任务就是确立在生存和成功适应的过程中各种心理状态和心理过程的功能。无需再质问"什么是心理的实质"，取而代之的问题变成"什么是心理活动的功能"。这也成为了威廉·詹姆斯心理学的核心原则——心理的永恒标准在于导向某个目的的行动。詹姆斯的学生桑代克（1874—1949）在他有名的"效果律"（Law of Effect）中也提到了相似的观点：行为的发生或多或少取决于行为所产生的效果。保证事情处于令人满意状态的功能变得尤为重要；而导致痛苦和受难的功能则越来越少。

　　查尔斯·达尔文(1809—1882)，曾在爱丁堡大学和剑桥大学研读医学和神学，然而这两门学科都无法激起他的兴趣；相反，他很早就开始就对一些自然现象感兴趣，并在"贝格尔号"上为自己争取到自然学家的研究职位（不付薪水的），这是一艘沿着南美洲太平洋沿岸进行探险的船只（1831—1836）。这一次旅程为他提供了许多数据，最终他将这些数据用于1859年出版的举世著作《物种起源》中。其间，他结婚并成为9个孩子的父亲，并且在科学界为自己赢得自然学家的称号,著书记录下在"贝格尔号"上的旅行，开始阅读托马斯·马尔萨斯的作品。《物种起源》专门研究非人类物种世界的进化，不过在《人类的由来》(1871)中他将进化论应用于人类。

神经心理学和神经学的进步

截至 1800 年，已经有证据说明肌肉受到一股电的刺激而活动，这个理论一开始受到质疑，但最终在 19 世纪早期开始受到肯定。1810 年（查尔斯·贝尔）到 1822 年（弗朗索瓦·马让迪）的解剖学研究显示，神经系统的感觉和运动功能在解剖学意义上是明显不同的。信息通过感觉神经从背侧面脊髓抵达大脑，而由大脑发出送到肌肉群的运动指令则通过腹侧面脊髓传递。

根据盖尔提出的理论——以及他的言论——19 世纪的研究认为大脑皮层、小脑、髓质、大脑深层是服务于特定的功能。保尔·布罗卡（,1824—1880）在对他的一位失语症患者的尸解中发现其左额叶有一处损伤，提出了"布罗卡氏失语症"（Broca's aphasia），认为人类心理发展的瞩目成就——语言——是一种基于大脑的能力。到了 19 世纪 30 年代，马歇尔·霍尔等人提出了一个反射功能的相关理论。在 19 世纪 40 年代，亥姆霍兹等人提供的实验数据显示神经传导速度——10 年前认为该速度几乎可以无限快——其实比较缓慢，只能达到每秒 20 到 40 米。不久之后，埃米尔·杜波依斯—雷慕（1818—1996）证实了神经系统中的电活动是一种化学反应，如今被称为离子反应过程。继而，神经临床和神经心理学实验室更加重视对生理心理学的研究，并且坚持认为"心理"仅仅是化学和物理反应过程的一个代号。这种观点在行为主义时期是"不折不扣"的谬论，这一点前文介绍中已经提到：将"心理"（生命、爱、品德等）仅仅理解为一种化学和物理过程。要接受这一观点则需一种乡农式的轻信——而且相信这一谬论的人不在少数。

行为主义

至少从19世纪至今，科学界中有个亘古不变的目标，就是证明科学的终极统一。这是恩斯特·马赫（1838—1916）在那个世纪末提出的，这也是他充满智慧的后代们——20世纪30年代的逻辑实证主义者的目标。尽管在该运动中，我们可以发现很多大人物之间存在实质性的意见分歧，但那时候（现在也是）在以下这些关键观点上普遍一致：第一，科学是一种独特的探究形式，最终的科学结论必须经由相关检测和观察来确定；第二，科学论题至少原则上是可观察性的，无论是直接观察，还是效果观察；第三，现实的终极"介质"是有形的——绝不存在鬼魂一说，或者是精神等其他；最后，物理学不是"形而上学"，不需要为神圣的目标或者隐形的设计而预留任何特殊空间。这种观点产生了心理学中的行为主义学派，约翰·华生在其作品中词彩飞扬地解释并支持这一学派。这也获得了条件反射研究的先锋伊万·巴甫洛夫的间接支持。后来斯金纳使其理论更为成熟并具有更大影响力。

进化论从单纯的自然进程意义上解释了生物的变异性和稳定性。某一物种的生物表现出自然变异，这些变异使某个个体有了更好的能力去迎接挑战，从而得以生存。最终，自然选择的特征在后代繁殖群中越来越常见，成功的"品种"成为了普通品种。一般来说，行为主义基于一个假设——以上相同的过程在适应行为意义上同样有效。那些带来积极结果的反应成为这个有机体全部能力的更为普遍的特征，那些不成功和不适应的行为则被"掐灭废除"。

伊万·巴甫洛夫(1849—1936)，俄罗斯生理学家，因研究消化而获得诺贝尔奖。他建立了条件反射原理，"条件"控制对特殊环境刺激的某种反映行为。他的研究证明，一种基本的生物反应——对放置于口中食物的唾液分泌反应——可以置于以前无效刺激之下来加以控制，如摇铃。频繁重复的"摇铃—食物"的连续过程可以使铃声引起唾液分泌反应，而不再需要食物。这就是著名的"经典条件反射"（又称为"巴甫洛夫条件反射"）。

伊万·巴甫洛夫（1849—1936），俄罗斯生理学家，诺贝尔奖获得者，提出了经典的条件反射理论。这些理论为当今神经生理学领域的"联想"理论奠定了基础。巴甫洛夫心理学是一种行为心理学，之后被完全不同的"斯金纳理论"（操作性条件反射）所取代。巴甫洛夫出生于俄国中部的一个名叫梁赞的乡村，他的父亲是一名东正教牧师。巴甫洛夫最初打算将来做牧师，但像达尔文一样——实际上深受达尔文的影响——他后来决定从事科学研究。他在彼得堡大学学习了化学和生理学，于1879年获得博士学位。之后，他并没有马上进行心理学研究；让他崭露头角的实验是有关消化系统的研究，这个研究主题为他赢得了1904年的诺贝尔奖，而对于精神病学这个新学科他则持怀疑态度。尽管他在"十月革命"后不时对苏联政府提出坦率的批评，但他的国际声望，和他作为科学进步的典范对苏联共产党具有的作用使他一直免遭迫害，直到87岁时去世。

巴甫洛夫还证明，一旦条件限制，一段连续的刺激可引起诸如唾液分泌一样的反应，这种刺激的连续统一性同建立条件反射所用的是一样。因此，一旦唾液分泌条件控制设定为5000赫兹的声调，1000赫兹、2000赫兹、6000赫兹等声调也将引起这种反应。在这样的条件下，巴甫洛夫观察到，当实验刺激渐渐不同于起先设置的条件刺激，唾液分泌量随之渐渐减少。这种普遍效应解释了所谓的"刺激类化"说。在差别条件作用时（比如，给食物时伴随一既定声调，而没有食物时则播放其他声调的声音）——条件反应精准地"调试"出对应于条件刺激的数值，这说明了"刺激分化"的过程。

约翰·华生（1878—1958），现代行为主义之父。他的演讲和著作几乎都是对"心理学为客观科学"的辩护，将可观察的行为作为学科的核心研究对象。他的文章《行为主义者视角下的心理学》于1913年发表在《心理学评论》，这篇文章向整整一代心理学家吹响了号角。他出生于美国卡罗来纳州格林维尔城外的一个农庄，在福尔曼大学获得本科和硕士学位。22岁的时候，他进入芝加哥大学研究生院学习，深受实用主义—功能主义教授们的影响，例如约翰·杜威、乔治·赫伯特·米德、詹姆斯·罗兰·安吉尔等，也是在那里他提出了行为主义理论。1908年，受聘为霍普金斯大学教授，5年后发动对主流心理学的攻击。华生是一个富有本能使命感的人，于1919年发表了著名的普及性教科书《行为主义心理学》，这本书对于推广他的理论具有重要的意义。正如他的弟子斯金纳，他希望自己的著作能够促使广泛的社会重组。1920年，他的妻子和他的前研究生学生提起离婚诉讼，他被迫从霍金斯大学辞职。最后，作为一个广告总监他在纽约度过余生。

巴甫洛夫推定,这些行为效应反映出动物大脑皮层的活动。他的理论基本上是一种条件反应的生理学理论。在美国,约翰·华生毫不留情地为行为心理学辩护,轻视由巴甫洛夫提出来的特殊心理过程,但同时又将巴甫洛夫的研究结果作为支持行为心理学的根据。行为主义是对可观察行为的研究,对于所谓的行为依赖的"心理"或"精神"的存在和运作不做任何假设。它的基本理论是科学的本质在于可观察性。我们可以观察他人的行为,而不是其意识。几乎毫无例外,行为主义与心灵主义是相对立的,后者是指将行为解释为心理活动和心理过程的结果的理论。

B. F. 斯金纳(1904—1990),弗洛伊德之外20世纪最具影响力的心理学家。他对老鼠和鸽子所做的操作性条件反射实验成为研究人类复杂行为的典型模型,其中包括打赌和赌博、孩子抚养、人际关系以及战争本身!在汉密尔顿大学主修文学之后,他到纽约开始了写作生涯,但以失败告终。之后他进入哈佛大学,1931年获得博士学位。1936年,他住在明尼苏达州大学,在约翰·华生的研究基础上,开始提出具有影响但又颇有争议的行为主义理论。他的小说《瓦尔登湖二号》(*Walden Two*, 1948)描述了一个建立在行为主义基础上的乌托邦社会。斯金纳丝毫不担心将自己的家庭成员置于这样的原则之下:他的二女儿在他的"婴儿箱"中度过了大部分婴儿期,这是一个在婴儿成长过程中能够由他控制环境影响的装置。在印第安纳大学做短期工作之后,他于1971年定居哈佛,在那里度过了自己余下的职业生涯。1971年,他发表了《超越自由与尊严》(*Beyond Freedom and Dignity*)一书,这是一套关于政治和社会思想的书,书名暗示了斯金纳心理学的原则。

因斯金纳具有影响力的著作，行为主义长久以来反对根据心理过程或者心理活动而产生适应性行为的解释。这是行为主义的一个中心定律，就像斯金纳所理解的行为主义那样，一种纯粹描述行为的科学没有义务去定位与适应性行为相关的内部过程和机制。行为的事实就在"那里"，可以观察到；其实际存在不会因为观察的层面不同（心理学的、基因的、解剖学的）而有所改变。为了符合这一观点，行为主义心理学家们确信，产生和控制行为的因素是有机体的外部环境，而不是生物学家所关心的内部环境。因此，对行为主义的批评往往将其冠以"空心有机体"心理学的名号。

神经心理学和认知神经科学

行为科学无需漠视内在状态和过程。从20世纪50年代开始,由于行为主义心理学家在很大程度上改善和提高了对行为的测试和控制,从而可能对认知、学习、动机、情感、社会关系等方面极其特定的生理的和生物化学的集层进行检测。在新千年中,心理学最具活力和最具发展潜力的领域可以说是认知神经科学。该学科包含了实验心理学、计算机科学、神经及神经生理学、人工智能和心智哲学。

认知神经科学的基础之一就是神经心理学。在过去这几十年中,这一研究的课题一直被归于不同的标题之下:生理心理学、心理生物学、生物心理学、医学心理学,而这几门学科之间难以明确区分。一般来说,生理心理学和医学心理学主要关注人类精神系统在健康和患病时的功能。生理心理学、心理生物学和生物心理学相对来说并非专门关注人类,而是更广泛地研究非人类的动物群体。不管归于何种学科标题之下,所有这些领域的研究和理论的共同之处是:都是研究心理过程或者性能或者能力,与特殊神经心理学、神经化学和神经解剖学之间的关系。

如上所述,这类观察和学说早在古希腊希波克拉底医学派出现之时就已存在。但经过几个世纪的研究,其间也经历了更长阶段的沉默期,才精确地解剖并且记录下人类和非人类的神经系统。不仅关于细节的研究一直等到17世纪显微镜的诞生,而且神经系统功能模式的发现也等至发明了可以记录电活动的仪器。直到20世纪，才有确凿的证据证明了结构独立的神经元以及神经系统的细胞单元确实存在。也是直到20世纪，才有证据表明信息从一个神经元传输到另一个是通过化学介质。

在20世纪初，就已有了神经元的专业解剖学证据；20世纪末，随着计算和可视化技术的发展，已能够实时展示整个大脑中不同结构的真实活动。同时，通过对临床神经学的系统研究，发展完善的"脑科学"通常能够在细胞层面识别那些与人类及其他生物心理中的感知、认知、行为、动机和情感等方面紧密联系的过程和活动。这也就是盖尔在18世纪末以他所能想象的方式发展起来的颅相学。

认知神经科学的出现是缓慢渐进式的。感觉和感知过程的基本研究为其奠定了基础，包括：感觉阈、注意力、信息处理、目标识别、短期记忆等研究。经过几十年的研究，心理功能模型被分为以下一系列模块：

1. 受到冲击的刺激导致感官产生反应。
2. 感官对这些刺激的特点，比如强度、大小和地点进行电子"编码"。
3. 这些编码后的信号被短暂存储于可以与之前存储信息相比较的地方。
4. 然后进行比较和评估，以决定新信息是否被保留还是被擦除。
5. 如果信息与情感和动机过程相关，那么这些信息就会被保留。
6. 来自储存记忆的信息会促进行为的发生。
7. 行为的即时结果将会通过反馈线路回到系统当中，因此可提供闭合回路（伺服的）机制使个体能适应不断变化的情况。

这是当动物和人类面对各种环境挑战时，假定会运行的某些"组件"的简单而不完全的描述。以这类模型为基础的理论，正从脑损伤效果、选择性刺激和特定大脑区域的记录以及对患有某种心理疾病的病人的尸解研究来寻找支持。

认知神经科学领域的主要代表人物都接受这类理论。复杂的心理过程被认为是来自更加基础的激活作用。因此"解决问题"这一术语涵括了一系列特殊功能：选择性注意，感觉编码，知觉注册，记忆，动机和情感的"增益"或放大，直觉或者习惯性行为调整的重获。很多人认为这些完全不同的功能依赖于大脑内部各种功能模块，这些模块以特定的渠道相连接，采取不同方式整合为一体来解决不同层级的问题。

尽管人们热情地提出这些理论，并为之辩护，但需要指出的是，至今人们还没有对其解剖结构——细胞集合——也就是"模块"的合成结构有一个清晰认识，也并不知道将这些细胞称为"记忆模块"是否合适。人类或者动物能够在迷宫里记住是向左转和向右转，这是清晰明了的。但要想清楚说明是大脑里某个结构"记住"的话，则不甚了了。这类描述背后的基本理论和动机来自"科学统一"的观点，用一种更加大众的说法，该观点寻求简化解释。其目的在于解释复杂事件时，将事件简化为更为简明或更为基本的事件，这种方式就像是解释水的特性：氢和氧以适当比例结合而成。它的支持者认为，这种方式的价值在于用客观的科学解释代替了主观模糊的说法。但是，仍然存在许多反对简化论的有力理由：

1. 将心理现象简化成心理事件的尝试可能无法得到解释，却可能导致减弱对该现象的兴趣。比如说，将颜色的体验简化为神经元和大脑细胞的放电过程，这将对带给这种体验可立刻感知的特质的研究毫无帮助。

2. 简化策略或多或少有可能弄巧成拙。毕竟能够让心理学家对于神经系统中发生的事件如此感兴趣是由于这些事件相关的可靠性，也就是心理事件和心理过程。简化这些也就是简化了心理学领域里神经系统的重要性。

3. 在不同的文本当中，人们理解对某个概念的解释大相径庭。在一起致命射杀中死去的受害者是（a）因为伤痛和失血,（b）因为袭击者知道他有一大笔钱,（c）因为这笔非法所得的钱是从袭击者那里偷走的,（d）因为袭击者正被暴徒勒索。但我们不能因为（a）是符合"客观科学原则"而肯定地说（a）就是最好的解释。

依然存在可信的理由，认为科学调查和解释的逻辑，就像物理学中建立的逻辑一样，最终不适用于解释构成心理学内容的一系列事件（文化、美学、道德、法律、政治和人际关系）。因此，简化策略不仅幼稚，而且最终会产生误导。

弗洛伊德和精神分析学

"弗洛伊德"心理学的广泛影响来自19世纪末一位在维也纳行医的临床精神病学家的工作,更清晰地了解某些"歇斯底里"症状的病因是他的主要目的。但这个事实常常被人们忽视。就像达尔文一样,弗洛伊德并没有打算去改变这个思想的世界。他在临床行医中发现,某些症状(残疾、失明、严重焦虑)与心理上令人不安的思绪或过去经历产生的压抑有某种关联。根据物理学中的守恒定律(这个定律在弗洛伊德的年代已经提出),似乎也存在一种守恒的心理能量,尽管它通过不同的方式表现出来。因此,歇斯底里症状可以被认为是一种心理压抑过程的物理表现。

1900年,弗洛伊德可以自由地从现今被科学界广为接受的进化论中借鉴理论。达尔文提供了强有力的论据来支撑论断:人的心理与非人类动物的心理具有相同的连续统。弗洛伊德接受生存本能促使人类行为这一观念,这种行为方式在动物世界中表现普遍。自然赋予动物能力,使其乐于寻求最终决定自身或该物种生存的活动。哺乳期的幼崽仅仅是因为感觉上的快感而不是营养上的考虑而接受哺乳。因此弗洛伊德提出了一个"心理性发育的理论",将动物由幼年到成年的性行为主要模式划分出阶段。终极阶段就是"异性生育性倾向",物种以此得以繁衍继续。

弗洛伊德心理学的中心理论要点就是"无意识动机"。人为其行为给出的原因很可能是由于受到无意识欲望的刺激，而如果将其公诸于世这种欲望会导致社会排斥。在社会化的过程中，一个正在长大的孩子自然会倾向于自我满足，否则就得服从"快乐原则"，面对由成人社会强加的"现实原则"。这些原则之间的较量会长达一生。完全受控于现实原则将会导致一种不太真实的生活；完全服从于快乐原则会使人无法适应社会。

因为欲望和行为源自所谓的无意识，一个人没有办法深刻且全面地理解为什么他或者她的目标、挫折、成功和失败会依次接踵而来。根据该理论，使成年人气馁的心理挫折最早在孩提时期就埋下了种子，直到成年依旧存在，弗洛伊德称之为"童

西格蒙德·弗洛伊德（1856—1939），一开始是执业神经医师，后来他遇到了患有"歇斯底里"症状的神经病人，这让他开始研究利用催眠术进行诊断和治疗。之后，他开始研究抑郁症，从而使他的精神抑郁理论成为整个研究心理疾病的"弗洛伊德"心理学的基础。弗洛伊德出生于摩拉维亚，4岁时举家迁到维也纳，一直到1937年被纳粹赶出。他不安定且不愉快的婚姻开始于1886年，同一年他建立起自己的私人诊所，治疗精神疾病。在弗洛伊德理论形成过程当中，他极大地受到法国神经学家让·夏柯以及他在维也纳的同事约瑟夫·布洛伊尔的影响。直到1908—1909年，第一次国际精神分析学会议召开，弗洛伊德到美国做了一系列的讲座，他才开始获得公众的关注。他是6个孩子的父亲，其中一个叫安娜的女儿后来也成为一个家喻户晓的精神分析学家。弗洛伊德于1939年在英国死于癌症。

年的痕迹"。只有通过深层分析，病人才能够回溯到种下祸根的那个时候。只有精神分析医师而不是病人，才能找到病人焦虑、希望、失败之中更深层次的意义以及重要性。

弗洛伊德的观点在表达上过于宽泛，因此至今都还没有一个清晰的概念。但是这一观点影响了哲学、文学批评、法律、艺术还有心理等各方面！现在看来，可能只有心理学所受影响没有其他方面那么大。至于其影响，或许最常见的有以下几方面：

1. 怀疑论。对于个人行为和文化价值合理解释的怀疑。根据弗洛伊德心理分析的理解，合理的解释是典型的"理性主义"，意在用有利的说辞来说明主动性和行为，这些主动性和行为实际上受达尔文种类说即所谓的兽性动机的驱使。达尔文＋弗洛伊德＝（大多数的）进化心理学和社会生物学。

2. 基本道德观念的相对论。心理学分析对"道德"的解释是一系列妥协让步：如果不可接受的责备和惩罚不可以避免，那么快乐原则必须让步于现实原则。道德范围里没有绝对，这种道德范围是由理性来决定，而且在任何情况下都有效；在这个范围里，只有自我与世界之间日常的冲突，原始欲望与压抑之间的冲突，原始欲望与象征转换之间的冲突。

3. 心理决定论。性心理发展阶段的恒定性、进化规律的不变本质以及文明所提出的要求，都作用于个人生活，使其成为一种不满足、满是冲突和无声绝望的生活。这一切都是"可能"的，也就是说，通过深度分析才能改善这种顺从的理解。

弗洛伊德自己期望心理学分析的条件和原则能够最终被重塑为发展完善的神经心理科学领域中的事实，这种期望受到各种肯定和质疑。心理分析理论只是第一步，他的终极目标是可以对精神进行生物学分析。但是，该理论基本上是一种叙述式的，几乎没有可检验性，一种可测量性——一言以蔽之，没有科学的认知性。这相当于一个叙述人类本质的故事；一个很多人觉得怪诞的恐怖故事，但也会有很多人认为这是真实的——"就如我生活的真实"。

　　对弗洛伊德理论的辩护和挑战已经花了足够的笔墨，就不再赘述双方观点了。根据临床医学家自己的说法，实际临床的心理疗法只是零散地得益于弗洛伊德，并没有从他的理论中得到"如何"诊断和治疗的方法。在弗洛伊德的理论中，达尔文的学说与达尔文自己的理论一样具有同等说服力，或者，就这点而言，在现今一些将这些原则视为真理的书籍或文章中，他的理论一样显得令人信服。第一次指导学生了解弗洛伊德理论时需要认识到：作为一个"理论"，它几乎没有达到科学理论所必需的要求；而且，适应于几乎与其理论相悖的任何证据正是弗洛伊德"理论"的本质所在。也许，最好的理解方法就是将其看成是对人类本质充满想象的解释，就像我们能够在一部小说里面发现的那样。

社会心理学

如果人们要认识心理学的三个主要"决定性"观点——行为主义、神经心理学和基因学——那么人们会很清楚地看到,所有的观点将会遭遇社会心理学的严峻挑战。有关旁观者效应、模拟监狱、同伴压力和服从的研究,都指出了行为和判断受到即时环境的影响。斯坦利·米尔格拉姆(1933—1984)做过一个著名且备受争议的实验,他要求参与者对受试者施加他们有理由认为会感觉痛苦甚至致命的电击,这个实验(错误地)被形容为"体罚对于学习和记忆行为的效用"。研究中有大约三分之二的参与者("教师")持续增加事先被告知的加在"学习者"身上的电压,实际上这些"学习者"是实验的合作者。当后者假装痛苦受难时,而实验室工作人员不需要给予"教师"多余的鼓励就能使他们遵从实验目的(继续施加强度更大的点击)。[1]菲利普·津巴多(1933—)在他的研究中也通过模拟监狱条件获得了对比实验效果。参与研究的斯坦福学生被随意分配为"囚犯"组或"警卫"组,身着警服的学生以其警卫职责需要维持囚犯的秩序。在实验的几天时间中,学生警卫对如今看来顺从

---

[1] 该实验又称为"米尔格拉姆实验",是一项关于权力服从的研究,其目的是为了测试参与者,在遭遇权威者下达违背良心的命令时,人性所能发挥的拒绝力量到底有多少——译者注。

的"囚犯"表现出虐待行为，这让许多囚犯宁可背叛其同伴以得到额外的毯子或者警卫的允许。[1]

此类研究的深刻启示意义是，人生可能会因为环境中强制压力的影响而放弃坚持保持体面的行为。但是这类发现也有个巨大的漏洞，那就是没有考虑到有一小部分可靠的群体，他们抵抗这种压力，并且保持对重要原则的忠诚。我们认为这类研究中真正有趣的是那些或多或少被忽略的小群体，因为通常注意力总是集中于那些从来都是逆来顺受的绝大部分人。

社会心理学实验往往能够高度反应一种现实特性，这种特性通常是实验室研究所缺乏的。社会心理学将研究对象置于真实的环境之中，或者提问他们在真实环境中会如何反应。这些环境的构成常带有明显的创造性，允许对偏见、自我评估、对一贯性的强烈倾向、对外界压力的易受害性等进行研究。在几十年的研究当中，关于人类本质又精心创造出另一个观点：一个人做出某种重要行为的源头和缘由很有可能与行为人自身给出的解释不同。尽管人类本质的概念事实上不是心理分析，但它还是再一次把注意力放在了超出个人意识控制的行为和知觉的决定因素上。当然，这一概念的趋向至少某些时候有悖于某些研究对象的趋向。而且，通常的问题还在于，这些预设的环境框架或者社会困境对于参与者限制的程度，远远超过他们在真实世界中受限制的程度。该研究和理论的重要性在于，它鲜活生动地提醒着，那些消除我们警觉而使人不知不觉从事一些活动的环境状态，如果经过深思熟虑我们就能规避这些活动。

---

[1]该实验又称为"斯坦福监狱实验"，是关于人类对囚禁的反应以及囚禁对监狱中的权威和被监管者行为影响的研究——译者注。

人类发展：道德和文明

在过去的半个世纪所做的具有想象力的和系统性的研究没有留下什么疑问，除了一点：婴儿一出生就具备了发达的感知能力，以及超过认知和解决问题的基本能力。同样清楚的是，从最早的发声阶段起幼儿实际上就在说话，也就是说，他们能够按基本规则发出声音——这种语言后来渐渐被成人语言所替代。确实，儿童是一个"小语言家"。

关于幼儿正常心理发育的育儿理论至今依然有很多争议。非人类动物的实验室研究虽然对此倾注了全部的注意力，但实际上并没有解决此类问题。也许这类研究中最具影响力的是哈利·哈洛（1905—1981）所做的项目，在实验中他将恒河猴幼崽与其母亲分离，只允许幼崽依偎一些用绒布或者铁丝做的玩偶。哈洛的研究得出：部分减少直至最少一个可供依偎的玩偶，这类分离所产生的可信后果是出现严重失调。[1]

毫无疑问，我们无法要求这类实验向聪明的人类说明，如果儿童从生命早期就脱离所有人类联系更有可能罹患行为适应不良，但是极端脱离产生的破坏性结果，也没有办法证明一个正常且蓬勃发育的个体所应必备的环境。厚厚的传记和自传文学已经足以证明，根本没有一个固定的模式。拥有优越关爱的

---

[1] 该实验被称为"代母实验"——译者注。

童年的人也会在后来的人生中一败涂地，而那些有过悲惨和虐待童年的人也能够名声大噪、生活丰足。因此孩子小时候该做什么不该做什么，并没有一个铁定不变的规律。然而，一个为知识所强化的常识是：对于那些值得我们尊敬和效仿的人，我们可以从他们人生的发展历程中找到许多成功必备的因素。

古希腊—罗马时期的哲学及心理学中有一个明确的思想：必须培养孩子成为一个公民（citizenship），这就要求对孩子进行系统的教育，而这种教育往往被翻译成"美德"教育。人们将希腊词"arête"翻译为"道德优尚"，是指具有一整套良好的行为和性情品格、自我控制力以及秉持高尚的人生目标。在该文明的顶峰时期，罗马制定了教育法，其中教育课程包括穿衣打扮、姿势仪态、说话声调、修辞手法等，当然还有"武艺"。很明显，对儿童文明生活方式的培养反映出了大文化环境中遵循的文明生活模式。罗马的存在不是为其公民，而是公民为罗马的光耀而活着。西方的民主以尊重个人尊严为基础，致力于保护基本的个人自由，因而需要一套不同的制度或教育，但是对于其中应该包含的内容至今没有统一的意见。为回答这些问题而进行的心理学研究也不多。

"道德教育心理学"出现时，竟然被划分为"认知心理学"领域里的一个小分支，在"社会心理学"领域里也只占相对像样的地位。认知心理学有着悠久的历史，但人们对它的研究兴趣时浓时淡，主要关注儿童怎么样理解道德问题，以及怎么得到解决办法。早期的研究来自让·皮亚杰（1896—1980），他采用讲故事的方法：一个男人的妻子病入膏肓，需要救治。最近的药店有药，但是药店所要价格很高，男人负担不起，而且药店坚持不付钱不给药。那个丈夫一直等到那天药店关门，然后破门而入，偷走了药，然后拿回家救自己的妻子。那么，那个丈夫做得对吗？对的话，为什么对？如果不对的话，又为什么不对？如果他被抓了，应该惩罚他吗？如此等等的问题。

皮亚杰发现，较小的孩子根据行为的后果来处理问题：如果要遭惩罚，那就是错的；如果能得到奖励，那就是对的。年纪稍大点的孩子或者年轻人，他们判断行为的道德性不是在乎别人怎么看待，而是根据个人所持的原则。由此，皮亚杰区分出两种认知阶段："他律"和"自律"。那些被他人或别种律则支配的人为他律阶段，而自我支配的就是自律阶段。几年之后，劳伦斯·科尔伯格（1927—1987) 在皮亚杰的基础上，通过研究来自不同文化的各个年龄阶段的人群，提出了道德发展的多阶段理论。该理论的一个原则是：人们必须连续不断地经历不同阶段，尽管阶段和年龄有关系，但是单纯的年龄并不是提高道德判断的保证。

如上所述，许多社会心理学研究都会考虑到人们无私行为的条件，欺骗行为的条件，因同伴压力或者迫于"处境"要求而破坏基本原则的条件。皮亚杰和科尔伯格的方法考虑产生道德判断的认知过程，然而社会心理学家则更倾向于考虑人们以某种方式判断和行动来达到社会目标。因此，一个人是否会帮助另一个处于困境中的人，在某种程度上取决于其他人是在场；若是在场，那么另外那个人是否愿意参与帮助。一个人如何从认知角度判断什么时候以及是否需要采取利他行为，与如果有这种需要表现出利他行为之间，这两者看上去有着明显不同。除了这些概述，当今的心理学几乎没有提供任何指导，甚至没有一个研究如文明发展这样核心问题的具体计划。

结　语：永恒的问题

我们可以清楚地看到，心理学作为一门研究和探索的学科可能涉及人类所有努力从事的活动，包括个人层面、社会集体层面，甚至整个国家组织层面。如果这门学科无法清楚地说明所有这些努力，我们也不能怪罪于心理学。作为实验科学领域的一门独立学科，心理学成立还不到两个世纪，还在继续完善它的方法，以及一定程度上仍在完善哪些议题是该学科的中心议题。现阶段的焦点还是关于方法论。也就是说，一些主要文章和学术机构多把心理学用于解决一些研究方法，而那些研究方法实质上是指出需要解决的问题。这是一种落后的发展方式。相反，应该先选择某种程度上对研究学科"正确"的问题和议题，然后再发展仅仅适合这些问题和议题的研究方法。

虽然致力于实验室研究并将统计模型用于描述和分析，心理学还是受困于对实际个体和个体心理的研究：这些个人想要实现个人目标，而且是出于一些离奇古怪的动机，还受到多样而易变的考虑因素的影响。诚然，目前公认的方法提供了可靠的数据统计描述，但这种采自集体的数据收集方式却洗掉了个人特性。心理学追求成为一门解密每个人的科学，却不能得出关乎任何人的结论。

在人类生活中，心理学表现最为生动的领域有三个：公民性（政治性）、审美性和抽象性。但这三个领域往往被忽略，大概是因为官方正式的"方法论"不能够适应这类事物。我们没有像政治心理学这样的学科；只有由下层人士组成的下议院的宣传机构，研究集中于，比如说投票行为。也没有美学心理学这样的学科；只有一些说明人们理解和欣赏什么作品，比如绘画和音乐。另外，虽然心理学一直以来都对"认知革命"议论纷纷，然而这个学科里却几乎鲜有探索抽象思维与诸如公正、公平原则之间关系的声音。

我们的道路还长，我们需要研究的还很多！

A STUDENT'S GUIDE TO
NATURAL SCIENCE

INTRODUCTION

The "natural sciences" include physics, astronomy, geology, chemistry, and biology, and are usually referred to as such in contradistinction to the "human sciences," such as anthropology, sociology, and linguistics. Of course, there is some overlap. Those disciplines which study human beings as biological organisms belong to both the natural and the human sciences.

In such a guide as this it would be impossible to give equal attention to all branches of natural science; I have therefore chosen to emphasize physics and, to a lesser extent, astronomy. There are several reasons for this choice. First, breakthroughs in these fields produced the Scientific Revolution and inaugurated the era of modern science. Second, physics can be regarded as the most fundamental branch of natural science, since the laws of physics govern the processes studied in all the other branches. Natural scientists tend to look at things from a "bottom up" perspective, in which the behavior of complex systems is accounted for in terms of the interactions of their constituents, and the branch of science that studies the most basic constituents of matter and their interactions is physics. Third, it can be said that developments in physics and astronomy have had the most profound impact on philosophical thought—along with Darwin's theory of evolution. Finally, there is the fact that I am myself a physicist.

Science was done in each of the great ancient civilizations of Asia, Africa, and the Americas. However, the story of science, as usually told, traces a path from the ancient Greeks and their precursors in Babylon and Egypt, through the Islamic world, and into Europe. There is good reason for this. All of modern science stems from the Scientific Revolution, which erupted in Europe in the 1600s and had its roots in the achievements of the ancient Greeks. The scientific developments that took place in other parts of the world in ancient times, though quite impressive in their own right, made little or no contribution to the Western Scientific Revolution and thus had hardly any lasting impact. (There are exceptions; for example, the concept of the number zero was first developed in India; it made its way into Europe through the Arabs.) From the sixteenth century through the nineteenth, advances in science came almost exclusively from within Europe's borders. It was not until the twentieth century that science became a truly global enterprise.

THE BIRTH OF SCIENCE

Natural science in the West was born in Greece approximately five centuries before the birth of Christ. It was conceived by the coming together of two great ideas. The first was that reason could be systematically employed to enlarge our understanding of reality. In this regard, one might say that the Greeks invented "theory." For instance, while literature is as old as writing, and politics as old as man, political theory and literary theory began with the Greeks. So too did the study of logic and the axiomatic development of mathematics. One of the earliest Greek philosophers, Heraclitus (540–480 B.C.), taught that the world was in constant flux, but that underlying all change is Reason, or *Logos*.

The second great idea was that events in the physical world can be given natural—as opposed to supernatural, or exclusively divine—explanations. The pioneer of this approach was Thales of Miletus (625–546 B.C.), who is said to have explained earthquakes by positing that the earth floated on water. He is most famous for speculating that water is the fundamental principle from which all things come. Thales was thus perhaps the first thinker to seek for the basic elements (or in his case, element) out of which everything is made. Others proposed different elements, and eventually the list grew to four: fire, water, earth, and air.

The search for the truly fundamental or "elementary" constituents of the world has continued to this day. In 1869, Mendeleev published his periodic table of chemical elements (which at that point numbered sixty-three). Later, the atoms identified by chemists were found to be composed of subatomic particles, which are now studied in the branch of science known as elementary particle physics. Today it is suspected that these particles are not truly elementary but are themselves manifestations of "superstrings." If this present speculation proves to be correct, it will vindicate Thales' intuition that there is but a single truly fundamental "stuff" of nature. In fact, as we shall see, this dream of theoretical unification and simplification has been progressively realized with each of the great advances of modern science.

The idea of "atoms" was the most remarkable and prescient of all the ancient Greek scientific ideas. It was proposed first by Leucippus (fifth century B.C.) and Democritus of Abdera (c. 460–370 B.C.). The Nobel laureate Richard Feynman, in his great *Lectures on Physics,* wrote,

> If, in some cataclysm, all of scientific knowledge were to be destroyed, and only one sentence were to be passed on to the next generation of creatures, what statement would contain the

most information in the fewest words? I believe it is the atomic hypothesis . . . that all things are made of atoms— little particles that move around in perpetual motion, attracting each other when they are a little distance apart, but repelling upon being squeezed into one another.

Of course, the rudimentary version of atomism proposed by Leucippus and Democritus was not a scientific theory in our modern sense. It could not be tested, and it led to no research program, but rather remained, as did most of Greek natural science, at the level of philosophical speculation.

The beginning stage of any branch of science involves simple observation and classification. Not surprisingly, much of Greek natural science consisted of this kind of activity. At times it was more ambitious and sought for causes and principles, but these principles were for the most part philosophical. In other words, they were not formulated into scientific laws in the modern sense. One thinks of Aristotle's principle that "nothing moves unless it is moved by another." This was meant as a general statement about cause and effect. It did not allow one to predict anything, let alone to make calculations.

It is interesting that the Greeks, for all their tremendous achievements in mathematics, did not go far in applying mathematics to their study of the physical world— astronomy being the major exception. This should not surprise us. It is perhaps obvious that the world is an orderly

Archimedes (c. 287–212 B.C.), one of the great figures of mathematical history, was born in Syracuse, Sicily. He discovered ways of computing the areas and volumes of curved figures, methods that were further developed by Torricelli, Cavalieri, Newton, and Leibniz in the seventeenth century in order to create the field of integral calculus. Unlike most Greek mathematicians of antiquity, Archimedes was deeply interested in physical problems. He was the first to understand the concept of "center of gravity." He also founded the field of hydrostatics, discovering that a floating body will displace its own weight of fluid, while a submerged body will displace its own volume. He used the latter principle to solve a problem given to him by King Hiero of Syracuse, namely, to determine (without melting it down) whether a certain crown was made from pure or adulterated gold. Hitting upon the solution while in the public baths, he ran naked through the streets shouting "Eureka!" ("I have found it!"), the eternal cry of the scientific discoverer.

Archimedes was also the discoverer of the principle of the lever, boasting, "Give me a place to stand and I will move the earth." Legend has it that he helped defend Syracuse from a Roman siege during the Second Punic War by inventing fantastic and ingenious weapons, such as the "claw of Archimedes," and huge focusing mirrors to ignite ships. According to Plutarch, "(Archimedes) being perpetually charmed by his familiar siren, that is, by his geometry, neglected to eat and drink and took no care of his person; . . . (he)was often carried by force to the baths, and when there would trace geometrical figures in the ashes of the fire, and with his fingers draw lines upon his body when it was anointed with oil, being in a state of great ecstasy and divinely possessed by his science." During the siege of Syracuse, in spite of standing orders from the Roman general that the great geometer not be harmed, Archimedes was struck down by a Roman soldier while drawing geometrical diagrams in the sand. His last words were, "Don't disturb my circles."

place, as opposed to being mere chaos; but the fact that its orderliness is mathematical is very far from being obvious, at least if one looks at things and events on the earth, where there is a great deal of irregularity and haphazardness. The first person to conceive the idea that mathematics is fundamental to understanding physical reality—rather than pertaining only to some ideal realm—was Pythagoras (c. 569–c. 475 B.C.). This insight was perhaps suggested to him by his research in music, where he discovered that harmonious tones are produced by strings whose lengths are in simple arithmetical ratios to each other. In any event, Pythagoras and his followers arrived at the idea that reality at its deepest level is mathematical. Indeed, Aristotle attributed to the Pythagoreans the idea that "things are numbers." This assertion may seem extreme, and doubtless did to Aristotle, but to the modern physicist it appears both profound and prophetic.

It is in the motions of the heavenly bodies that the mathematical orderliness of the universe is most apparent. This has to do with a number of circumstances. First, interplanetary space is nearly a vacuum, which means that the movements of the solar system's various bodies are unimpeded by friction. Second, the mutual gravitation of the planets is small compared to their attraction to the sun, a fact that greatly simplifies their motion. In other words, in the solar system nature has provided us with a dynamical system that is relatively simple to analyze. This was vital for the emergence of science. In empirical science it is important to be able to isolate specific causes and effects so that they are not obscured or disrupted by extraneous and irrelevant factors. This normally has to be done by conducting "controlled" experiments (experiments, for instance, that allow one to compare two systems that differ in only one respect). Usually, it is only in this way that one gets a chance to observe interesting and significant patterns in the data. But it did not occur to the ancient Greeks to perform controlled experiments—or, for the most part, experiments of any kind. It is therefore very fortunate that they had the solar system to observe.

The first application of geometry to astronomy seems to have been inspired by Pythagorean ideals. Pythagoras himself suggested that the earth is a perfect sphere. Later, Eudoxus (c. 408–c. 355 B.C.) proposed a model in which the apparently complex movements of the heavenly bodies resulted from their motion in perfectly circular paths. These Pythagorean principles—that theories should be mathematically beautiful and that they should explain complex effects in terms of simple causes—have been tremendously fruitful in the history of science. But they are not sufficient. The mathematical approach of these earlier astronomers lacked a key ingredient, namely the making of precise measurements and the basing of one's theories upon those measurements. In this respect Hipparchus (c. 190–c. 120 B.C.) far exceeded his predecessors and transformed astronomy into a quantitative and predictive science. He made remarkably accurate determinations of such quantities as the distance of the moon from the earth and the rate at which the earth's axis precesses (the so-called precession of the equinox, a phenomenon that he discovered). After Hipparchus, the development of ancient Greek astronomy reached its culmination in the work of Ptolemy (c. 85–c. 165), whose intricate geocentric model of the solar system was to be generally accepted for the next fifteen centuries.

The Greeks' attraction to mathematics was a double-edged sword. On the one hand, it had incalculable benefits for science. The Greeks' most enduring scientific legacy lay in their mathematics and mathematical astronomy. On the other hand, this attraction to mathematics reflected a tendency (seen very markedly in Plato) to disdain the world of phenomena for a more exclusive focus on the realm of the ideal.

We find quite the opposite tendency in Aristotle. Aristotle was very much interested in phenomena of every kind, and (in contrast to many of his later epigones) engaged in extensive empirical investigations, especially in biology. He was undoubtedly one of the greatest biologists of the ancient world. His legacy in physics, however, is more ambiguous—indeed, on the whole, perhaps negative. There are several reasons for his comparative failure as a physicist. First, there is the fact, already remarked on, that terrestrial phenomena are hard to sort out. Among many other complications, they involve large frictional forces, which resulted in Aristotle being fundamentally misled about the relationship between force and motion. Second, Aristotle appreciated neither the true nature of mathematics nor its profound importance; his genius lay elsewhere. Third, Aristotle's approach to the physical sciences was philosophical; in his thought there is no bright line between physical and metaphysical concepts. This would not have created so many problems for later thought—problems discussed in more detail below—had it not been for the very brilliance and depth of Aristotelian philosophy.

Hipparchus (c. 190–120 B.C.) is considered the greatest observational astronomer of antiquity. Little is known of his life except that he was born in Nicaea, located in present-day Turkey, and spent most of his life on the island of Rhodes. What distinguished him from his predecessors was his application of precise measurements to geometrical models of astronomy. Not only did he make extensive measurements; he also made use of the voluminous astronomical records of the Babylonians, which dated back to the eighth century B.C. This long span of data allowed him to compute certain quantities with unprecedented accuracy.

Hipparchus created the first trigonometric tables, which greatly facilitated astronomical calculations, and developed or improved devices for astronomical observation. He also compiled the first star catalogue, which gave the positions of about one thousand stars. Though he worked on many problems, such as determining the distance to the moon, he is most famous for discovering the "precession of the equinox" and correctly attributing it to a wobbling of the earth's axis of rotation. Newton later showed that this wobbling was caused by the gravitational torque exerted by the sun and moon on the earth's equatorial bulge.

THE SECOND BIRTH OF SCIENCE

It is sometimes said, by those with an axe to grind against religion, that the rise of Christianity brought an end to the first great age of scientific progress. This claim is untenable. It is true that one can find statements in the writings of the church fathers that deprecate the study of nature, and that science was not high on the early Christians' list of concerns. However, one finds the same range of attitudes toward science among the early Christians as among their pagan contemporaries. And the fact is that the glory days of ancient science were long gone by the time Christians became a significant demographic or intellectual force. The golden age of Greek mathematics ended two hundred years before the birth of Christ. (For example, the great Greek mathematicians Archimedes, Eratosthenes, and Apollonius of Perga died, respectively, in 212 B.C., 194 B.C., and 190 B.C..) Only a few great figures in ancient Greek science date from the period after Christ, notably the astronomer Ptolemy, who died around A.D. 165, and the mathematician Diophantus, who died around A.D. 284. At that point, Christianity was still a small and persecuted sect.

As is well known, an impressive revival of mathematics

and science began in the Islamic world in the ninth century. Under the Abbasid caliphate, which stretched from North Africa to Central Asia, scholars were able to draw upon the patrimony of the Babylonians and Indians as well as the Greeks. The Muslim contributions to science are memorialized in the many scientific terms of Arabic origin, such as *alcohol* and *alkali* in chemistry (a field of inquiry once called "alchemy"); *algebra, algorithm, and zero* in mathematics; and *almanac, azimuth, zenith,* and the names of the bright stars *Algol, Aldebaran, Betelgeuse, Rigel,* and *Vega* in astronomy. However, the brilliance of Muslim science began to fade after a few centuries. The Islamic theological establishment tended to be indifferent or hostile to speculative Greek thought, and therefore science did not achieve the kind of institutional status in the Muslim world that it later achieved in the universities of Europe.

The second birth of science really came in the Latin West. In the eleventh century, when Western Europe began to recover from the economic and cultural collapse caused by the barbarian invasions, its scholars became aware again, largely through contact with the Arab world, of the ancient Greeks' great achievements in science. This awareness engendered an insatiable curiosity about and demand for the works of ancient Greek scholars, which led in turn to a frenzy of translations of these works into Latin, either from Arabic sources obtained in Spain or directly from Greek versions obtained from the Byzantines. Universities were invented in medieval Europe, and they were founded in part as places where this newly recovered knowledge could be studied. The intense interest in Greek science—or, as it was called at that time, "natural philosophy"—was shared by clergy and laity alike. Indeed, in medieval universities the study of natural philosophy was a prerequisite for the study of theology. (This would be somewhat analogous to physics being a required course in today's seminaries.)

For a long time, it was standard for modern scholars to dismiss medieval science as lacking in creativity or true scientific spirit, and as being quite irrelevant to later scientific progress. However, scholars such as Pierre Duhem and A. C. Crombie successfully challenged that consensus. They demonstrated that medieval science was far more vital than had been supposed, and that the picture of monkish scholars slavishly following Aristotle had been overdrawn. The "natural philosophers" of the Middle Ages were quite aware of some of the inadequacies in Aristotle's ideas and adopted a cautiously critical approach to him, though their interesting critiques were not based on experiments but on logical reasoning—and to some extent on what we would today call "thought experiments." In addition, the medievals took tentative steps toward developing a science of motion. The crucial concept of uniform acceleration (or in their quaint terminology, "uniformly difform" motion) became understood for the first time; the notion of "impetus" (an anticipation of the concept of "momentum") was developed;

graphs were invented to facilitate reasoning about mathematical functions and motion; and mathematical laws of motion were first proposed. Some historians, such as Duhem, Crombie, and more recently Stanley Jaki, have even claimed that these ideas directly influenced the thinking of Galileo and other founders of the Scientific Revolution (though the extent of this influence is disputed, and the issue is far from settled).

Be that as it may, there is one way in which the revival of science in medieval Europe certainly did lay the groundwork for the Scientific Revolution to come. It "institutionalized" science, as Edward Grant, the noted historian of science, has put it. In the ancient and Arab worlds, science like art had depended upon the patronage of wealthy or powerful individuals who happened to have a personal interest in it. It was therefore a hit-or-miss affair, subject to the vicissitudes of politics and economic fortunes. By contrast, in the medieval universities there was created for the first time a stable community of scholars that studied scientific questions continuously from generation to generation. That is, a scientific community came into being. By the end of the Middle Ages there were nearly one hundred universities in Europe, and their graduates numbered in the tens of thousands. This created a significant literate public that was interested in science, was willing to pay to be taught or obtain books about it, and from whose ranks scientific talent could emerge.

Without the scientific community and the scientific public created by the medieval universities, the Scientific Revolution would not have had fertile soil in which to germinate.

SCIENCE, RELIGION, AND ARISTOTLE

The foregoing discussion raises two very interesting and difficult questions: Why did the Scientific Revolution occur in Europe, and what was the role of religion in that revolution? One view, encountered so frequently that it has become a cliché, is that the Christian religion was the enemy of science and tried to strangle it at its birth; this animus is alleged to have been clearly revealed in the Galileo affair. However, scholars no longer take this view seriously. The idea that the church establishment has been implacably hostile to science is a myth created to serve the purposes of antireligious and anticlerical propaganda.

In fact, the church has always esteemed scientific research—even at the time of the Galileo affair. We have seen that the medieval church was willing to embrace the science of the ancient Greeks, even though it was naturalistic in character and pagan in origin, and that this science had an important place in the curricula of medieval universities, institutions that had been founded primarily under church auspices, received much church patronage and protection, and were staffed largely by clerics. Indeed, most of the scientists of the Middle Ages were clergymen, such as

Nicholas Oresme (c. 1323–82), who was bishop of Lisieux and a mathematician and physicist of great ability. This tradition of clergy involvement in scientific research has continued to the present day. In fact, from the seventeenth through the twentieth century, a remarkable number of important scientific contributions have been made by Catholic priests.

The favorable attitude of the church toward "natural philosophy" was largely the result of the efforts of St. Albert the Great, who helped introduce Greek science into the medieval universities, and his pupil St. Thomas Aquinas. Both men were convinced of the possibility and importance of harmonizing faith and reason and saw in the philosophy of Aristotle the conceptual tools needed to accomplish that. This gave a tremendous impetus to the study of natural philosophy and thereby helped prepare the way for the Scientific Revolution. However, the church's embrace of Aristotle also had negative consequences, since it helped to cement in place a mistaken approach to the physical sciences.

The philosophical system of Aristotle, as transformed and Christianized by St. Thomas Aquinas, was a brilliant and impressive intellectual synthesis. One can think of it as a medieval "grand unified theory" or "theory of everything," except that it comprised not only natural philosophy, but also metaphysics, anthropology, moral philosophy, and much else, including even truths of revealed religion (insofar as they could be grasped by human reason). This Aristotelian-

Oresme, Nicholas (c. 1323–82) was born near Caen, Normandy, received his doctorate in theology from the University of Paris, became counselor and chaplain to the king of France, and was ultimately installed as the bishop of Lisieux. A polymathic genius, he made pioneering contributions in mathematics, physics, musicology, and psychology. He was an important figure in scholastic philosophy, and is considered by many to be the greatest economist of the Middle Ages. In mathematics, he was the first to discuss fractional and even irrational exponents, and his studies of infinite series proved that the harmonic series (1 + 1/2 + 1/3 + 1/4 + . . .) diverges. Oresme was also the first to use graphs to plot one quantity as a function of another, extending his discussion to three-dimensional graphs, thus anticipating by three centuries some of the key ideas of Cartesian analytic geometry. He used such graphical methods to give the first proof of the "Merton theorem," which provides the distance traversed by a uniformly accelerated body. These studies probably contributed indirectly to Galileo's discovery of the law of falling bodies.

In addition to all this, Oresme proved that phenomena could be accounted for satisfactorily by assuming that the earth rotates rather than the heavens. The analysis by which he refuted common physical objections to this was superior to those later articulated by Copernicus and Galileo, especially in its resolution of motion into vertical and horizontal components. He was the first to understand the distortion of the apparent positions of celestial objects near the horizon as resulting from the bending of light passing through air of varying density. And he argued for the existence of an infinite universe, contrary to the standard Aristotelian view. All in all, he must be accounted one of the great original thinkers in the history of mathematics and physics—and an important forerunner of the Scientific Revolution.

Thomistic philosophy has great strengths and remains indeed very much a living tradition. However, for all the light Aristotle's ideas shed upon metaphysical and moral issues, they contained much that was deeply mistaken and misleading when it came to physics and astronomy.

It is foolish to fault the medievals for adopting Aristotle's physics, for there was nothing else around; it was the science that they inherited from the Greeks. And for a time it did play the useful function (as even wrong theories can) of providing a framework for theoretical discussion and analysis. It also provided an example of a naturalistic theory of physical phenomena, which is no small thing. Nevertheless, it had the effect of leading scientific thought into a cul-de-sac from which it took considerable effort to escape. To the extent that theology contributed to prolonging the dominance of Aristotelian natural philosophy, it played an unhelpful role.

The dominance of Aristotelianism helps to explain the church's condemnation of Galileo and heliocentric astronomy in 1633. However, it was only one factor in a very complex affair. The other reasons for Galileo's condemnation included professional rivalry, Galileo's talent for making enemies, and, most important of all, the turbulence of the times. It was an era of great religious tension; Europe was being torn apart by the Thirty Years War, which had begun as a Catholic-Protestant struggle. As part of its effort to defend itself against the Protestant challenge, the Catholic Church had enacted at the Council of Trent (1545–63) a set of rules for the interpretation of scripture

---

Galilei, Galileo (1564–1642) was born in Pisa and began studies at the University of Pisa in 1581. He secured a professorship there in 1589, but decided to move to the University of Padua two years later because of conflicts with Aristotelians. In 1609, having heard of the invention of the telescope, he devised his own and with it began studying the heavens. His discoveries of sunspots, the moons of Jupiter, mountains on the moon, and the phases of Venus undermined Aristotelian science, refuted the Ptolemaic system, and made him a celebrity. He aroused opposition by his advocacy of the heliocentric Copernican system, and in 1616 the Roman Inquisition issued an injunction forbidding him to defend Copernicanism "in any way." At the same time, all books were prohibited that advocated Copernicanism as true rather than merely as a "hypothesis" (by which was meant a mathematical device for simplifying calculations). In 1623, Maffeo Cardinal Barberini, a friend and protector of Galileo, was elected pope. Unaware of the 1616 injunction, he did not object to Galileo defending Copernicanism as a "hypothesis." Galileo proceeded to publish in 1630 his *Dialogue Concerning the Two Chief World Systems*, in which he not only defended Copernicanism as true, but seemed also to lampoon the pope's philosophical opinions. The pope was outraged at this betrayal by someone he had protected; the forgotten 1616 injunction was discovered in the files; and in 1633 Galileo was forced to publicly renounce Copernicanism and sentenced to lifelong house arrest, which he served in his villa in Florence, where he was allowed to receive visitors and publish on other scientific subjects. In 1638 he published his *Dialogues Concerning Two New Sciences,* which set out his discoveries in physics, his greatest contributions to science.

that was intended to prevent radical theological innovations. Though reasonable in themselves, these rules ended up being misapplied to Galileo, who had unwisely allowed himself to be drawn into scriptural and theological debate by his enemies.

The condemnation of Galileo, though a fateful blunder, was not the result of hostility toward science on the part of church authorities, nor did it reflect an unyielding dogmatism in scientific matters. The words of Cardinal Bellarmine, head of the Roman Inquisition at the time of Galileo's first encounter with it in 1616, are well worth remembering:

> If it were demonstrated that the sun was really motionless and the earth was in motion] we should have to proceed with caution in interpreting passages of Scripture that appear to teach the contrary, and rather admit that we do not understand them than declare something false which has been proven to be true.

Bellarmine went on to say that he had "grave doubts" that such a proof existed and that "in case of doubt" one must stay with the traditional interpretation of scripture. As a matter of fact, such a proof of heliocentrism did not exist in Galileo's time, though there were strong indications in its favor for those with eyes to see them.

In any event, if we look at the eight-hundred-year record of the church's involvement with science, it is hard to see the Galileo affair as anything but an aberration. Far from seeing religion as an obstacle to the emergence of modern science, some scholars have argued that Christian beliefs played a part in making the Scientific Revolution possible. A great deal can be said for this view. For example, an idea fundamental to all science is that there exists a natural order. That is, not only is the world orderly, but it is in fact a *natural* world. We have seen that this idea arose among the pagan Greek philosophers, but Judaism and Christianity also helped promote this way of thinking. Whereas in primitive pagan religion the world was imbued with supernatural and occult forces and populated by myriad deities— gods of war, gods of the ocean and of the earth, goddesses of sex and fertility, and so forth—Jews and Christians taught that there was only one God who was to be sought not *within* nature and its phenomena and forces, but outside of nature, a God who was indeed the author of nature. In this way, biblical religion desacralized and depersonalized the world. To borrow Max Weber's term, it "disenchanted" the world.

For example, the book of Genesis, which is often seen as an instance of primitive mythmaking, was actually written in part, scholars tell us, as a polemic against pagan supernaturalism and superstition. When Genesis says that the sun and moon are merely "lamps" placed by God in the heavens to light the day and night, it is attacking the pagan religions that worshipped the sun and moon. When it says that man is made "in the image of God" and is to exercise "dominion" over the animals, Genesis is, among other things, attacking the paganism in which men worshipped and bowed down to animals or to gods made in the image of animals.

Medieval Christians were so comfortable with a naturalistic view of the physical world that it was commonplace as early as the twelfth century, according to Edward Grant, for philosophers and theologians to refer to the universe as a *machina*, a "machine." (Of course, both Judaism and Christianity teach the possibility of miracles. But miracles—from the Latin *mirari*, "to wonder at"—are such precisely because they are rare occurrences that contravene what the medieval philosophers called the "course of nature." They derive their whole significance from the fact that there is a natural order that only God, as author of nature, can override.)

A second idea that Judaism and Christianity likely helped to foster is that the universe is not merely orderly but *lawful*. A Christian writer of the second century, Minucius Felix, wrote:

> If upon entering some home you saw that everything there was well-tended, neat, and decorative, you would believe that some master was in charge of it, and that he was himself much superior to those good things. So too in the home of this world, when you see providence, order, and law in the heavens and on earth, believe that there is a Lord and Author of the universe, more beautiful than the stars themselves and the various parts of the whole world.

This emphasis on law goes back to the Hebrew scriptures, the first five books of which are indeed called by Jews the *Torah* or "Law." Of course, Israel regarded

---

Stensen, Niels (1638–86, also called Steno) made fundamental contributions to anatomy, geology, paleontology, and crystallography. Born in Denmark, he enrolled in the University of Copenhagen to study medicine. In his twenties he was already recognized as one of the leading anatomists in Europe. His anatomical studies greatly increased knowledge of the glandular-lymphatic system. (Stensen's duct, Stensen's foramina, and Stensen's gland are all named after him.) He also did important research on heart and muscle structure, brain anatomy, and embryology. After spending time in Paris and the Netherlands, he traveled to Florence to join the Accademia del Cimento, which had been founded by Ferdinando d'Medici, Grand Duke of Tuscany, to carry out experimental research in the tradition of Galileo, some of whose students were among its members. While dissecting the head of a great white shark that had been caught off Livorno, Stensen realized that the teeth bore a remarkable resemblance to the "tongue stones" found in great abundance in Malta. This led him to undertake those investigations for which he is now regarded as the founder of the scientific study of fossils and of the branch of geology called stratigraphy. His ideas on how geological strata form played a crucial role in unlocking the history of the earth and eventually in revealing the planet's great age. In addition, Stensen discovered the basic law of crystallography, which is that the angles of a given mineral are always the same ("Steno's law"). Stensen converted to Catholicism in 1667, was ordained a priest in 1675, and became a bishop two years later. As a bishop he was given to rigorous asceticism and was an ardent champion of the poor. He was beatified by Pope John Paul II in 1988.

God as its lawgiver, but he is also thought of as lawgiver to the cosmos itself. God says in the book of Jeremiah, "When I have no covenant with day and night, and have given no laws to heaven and earth, then too will I reject the descendants of Jacob and of my servant David." Psalm 148 tells of the sun, moon, stars, and heavens obeying a divinely given "law that will not pass away." The ancient rabbis said, "The Holy One, blessed be he, consulted the Torah when he created the world." The Torah was thus a law that existed eternally in the mind of God and according to which the universe itself was made. (This notion of an eternally preexisting law became linked in Jewish thought to the idea of the divine Wisdom, personified in later books of the Old Testament, especially Proverbs, Wisdom, Sirach, and Baruch. This divine Wisdom became in the New Testament the divine *Logos*, which means "Word" or "Reason." In other words, the idea that reason underlies the universe emerged within both pagan Greek and Jewish thought.)

It is not only religious authors who have seen the idea of a divine lawgiver as a factor contributing to the emergence of modern science. In explaining why Chinese civilization, with all its refinement and splendid achievements, did not produce a Newton or a Descartes, the decidedly nonreligious biologist E. O. Wilson pointed to the fact that Chinese scholars had abandoned the idea of a supreme being with personal and creative properties. No rational Author of Nature existed in their universe; consequently the objects they meticulously described did not follow universal principles. . . . In the absence of a compelling need for the notion of general laws— thoughts in the mind of God, so to speak—little or no search was made for them.

The cosmologist Andrei Linde, himself an atheist, has also suggested that the idea of a universe "governed by a single law in all its parts" has some connection to monotheism.

On the other hand, of itself monotheism is not enough to produce a scientific revolution. Islam is monotheistic, but the progress of science in Muslim lands eventually petered out. And Byzantine civilization, even though it was Christian and had never forgotten the science of the ancient Greeks, produced little science of its own. It may well be that the especially strong emphasis in the theology of the Western church on law, system, and reason contributed decisively to the emergence of modern science.

But something else was also at work during the Renaissance: a spirit of innovation, restlessness, and questioning that had to see things for itself and was skeptical of received opinion. For some, this even extended to theological matters—although, contrary to what many imagine, religious skepticism does not appear to have been a generative factor in the Scientific Revolution. Almost all of the great founders of modern science, including Copernicus, Kepler, Galileo, Boyle, Hooke, and Newton, were religiously devout. Some of them, like Kepler and Boyle, were clearly inspired to do scientific research by their religious beliefs whereas none of them were motivated by opposition to faith or orthodoxy. Until at least the mid-nineteenth century, most great scientists were religious believers; many continue to be so today.

THE SCIENTIFIC REVOLUTION

THE SCIENTIFIC METHOD

The Scientific Revolution was characterized by three great achievements. First, there was the shattering of the Aristotelian synthesis and a decisive break with its philosophical, speculative, and qualitative approach to doing science (even though science continued to be referred to as "natural philosophy"). Second, there was the realization of the importance of doing experiments and making precise measurements. This involved the growing use of artificial devices such as telescopes, pendulums, and vacuum pumps in investigating nature. (This second achievement also required a conceptual advance; in Aristotelian thought, machines had been regarded as causing things to move contrary to their "natural" motion.) And, finally, there was the mathematization of science. Of course, science had previously employed mathematics and even experiment, but it was in the 1600s that these tools came together to create a powerful new way of investigating the world, often called "the scientific method."

The scientific method comprised (a) the collection of

data by precise measurements and controlled, repeatable experiments, (b) the formulation of testable hypotheses to explain either regularities or anomalies in the data, and (c) the verification or falsification of those hypotheses by comparing their predictions with the results of new measurements or experiments. Unfortunately, the scientific method is sometimes spoken of as if it were an automatic or mechanical process. It is not a *process*, but an *activity*. Processes can be undertaken by machines. Activities require the imagination, insight, cleverness, initiative, creativity, and judgment of persons.

It is not sufficiently stressed in scientific education, where the main object is to master an enormous body of fact and theory, just how marvelously ingenious experiments and observations can be. Great experiments, like great theoretical ideas, are things of beauty. However, they tend to be more ephemeral. They are the scaffolding in the construction of the edifice of science and are too often forgotten after they have played their part.

Many also have an unrealistic understanding of scientific theorizing. Students are usually presented with a theory as a finished product, with all obscurities removed, essential ideas crystallized into precise concepts, fundamental principles identified, and logical structure clarified. This is indeed the best way for students to master the ideas. However, it can also lead them to lose sight of the messy, confusing, and arduous struggle by which the key insights were originally won.

A strong case can be made that a lack of appreciation of the actual methods of science has had harmful effects on philosophy. The two great contrary movements in European philosophy in the seventeenth and eighteenth centuries—rationalism and empiricism—were both to some extent inspired by the progress of science and considered themselves scientific in spirit. However, each exaggerated one pole of scientific thought at the expense of the other. The rationalists tended to think of all knowledge as advancing by a process of deductive reasoning from first principles, as in mathematics. They undervalued the empirical component of scientific progress. On the other hand, the empiricists underappreciated the role of hypothesis, abstraction, and theoretical construction in advancing scientific knowledge. Like the rationalists, they may also have been misled by a false analogy with mathematics. In mathematics, each conclusion must be firmly demonstrated before it can be used as the basis for further reasoning. Some have imagined that empirical science works in a similar way. They think that the existence of every entity posited by a theory and the truth of every theoretical proposition must be directly verified before the next step is taken. That is not how things work.

Take, for example, the theory of electromagnetism formulated by James Clerk Maxwell in the mid-nineteenth century. It posits the existence at every point in space and time of a three-component "electric field" and a three-component "magnetic field." No one has ever directly verified the existence of such entities everywhere in space-time, nor would it be possible to do so. However, it is not necessary. That is not how the validity of Maxwell's theory was established in the first place or why physicists still believe it to be true. Maxwell's theory, like any sophisticated scientific theory, is a highly elaborate and abstract structure that presupposes the existence of many things, not all of which can be directly or separately observed. Rather, it is the *theory as a whole* that is verified on the basis of observations; and these observations, however numerous they may be, are necessarily few compared to the entities that the theory assumes to exist.

The empiricists seemed to think that humans built up a picture of reality by adding together a large number of sensory "impressions," from which more complex ideas got generated by a process of "association." However, one does not directly sense magnetic fields (unless one is a monarch butterfly, say). One infers their existence by their often very indirect effects, and even then only with the help of abstract theory. And yet these magnetic fields are as real and as physical as rocks and trees. (The same point is illustrated by the electromagnetic spectrum: we directly sense "visible light" with our eyes, but can only infer the reality of ultraviolet light and radio waves by indirect means. However, this difference is due simply to the characteristics of our sensory organs. Radio waves, ultraviolet light, and visible light are in themselves equally real, differing only in wavelength.) The crude notion of verification that one finds implicitly in the

British empiricists of the eighteenth century also afflicted later versions of empiricism, such as the early-twentieth-century philosophical school called "logical positivism." For the logical positivists, every meaningful statement had to be translatable into statements about sensory impressions.

The relationship between theory and observation in science is complex and dynamic. Theories are built on experiment, and experiments depend for their interpretation upon theory. This "theory-dependence of experiment" is much talked about in recent times and has provided an opening for some "postmodern" thinkers to claim that the scientific method involves a "vicious circle" that somehow vitiates the notion of scientific objectivity. One can see through such sophistry by a simple analogy: maps were made by explorers; and explorers had to make use of existing maps. This "circularity" obviously did not prevent better and better maps from being made, nor does the dynamic interaction of theory and experiment prevent better and better theories of the physical world from being made. Indeed, that is precisely how they must be made; and the recognition of this fact was the fundamental achievement of the Scientific Revolution of the seventeenth century.

Before we turn to the history of that revolution, it is worth saying a bit more about how scientific theories are verified. In mathematics a theorem may be proven rigorously, in which case it can be affirmed with certainty; or it can be disproved, in which case it can be denied with certainty. However, in natural science, as in life, one accepts a theory with a degree of confidence that is, as it were, a continuous variable. In some cases, the confidence may be so great that we can speak of virtual certainty—indeed, of a "scientific fact." For instance, no scientist seriously doubts anymore that matter is made of atoms or that the earth rotates on its axis. In other cases, the confidence of scientists in a theory may be fairly strong, but not so strong that they would say it has been "confirmed." In yet other cases, there exists simply what lawyers would call a "rebuttable presumption" in favor of a theory. It all depends on the quantity and type of evidence.

What counts as evidence for a theory? It is not only a matter of quantitative predictions confirmed by experiment. For one thing, a particular experimental number, or even many such numbers, might be accounted for in several ways. For example, one of the major successes of Einstein's theory of gravity (the so-called general theory of relativity) was that it predicted accurately the "precession of the perihelion of Mercury" (the slow shift over time of the point of closest approach of Mercury to the sun). However, one could have accounted for the perihelion shift in several other ways. One way was to posit a certain amount of solar oblateness (i.e., a flattening at the sun's poles). Another was to posit a certain density of matter filling the space between Mercury and the sun. Nowadays, enough is known about the sun and its environment that these alternative explanations are no longer viable. But it is possible to find other alternatives; one could simply add, for instance, a new term to Newton's inverse-square law of gravity to account for the perihelion shift. So the successful prediction of that shift, though vitally important, was not the only reason that physicists began to believe in Einstein's theory.

Many considerations influence scientists' judgments about the plausibility or likelihood of a theory. These include the theory's *simplicity and economy*, its provision of a *more unified and coherent picture* of nature, its *explanatory power*, its *mathematical beauty*, its grounding in *deep principles*, its prediction of *new phenomena*, and its ability to *resolve theoretical puzzles or contradictions*. Einstein's theory of gravity, for instance, resolved the problem that Newton's theory of gravity was not consistent with the principles of the special theory of relativity. It also explained why "inertial mass" is equal to "gravitational mass." It predicted the phenomenon of the bending of rays of light by gravity. It flowed from the deep "equivalence principle." It was based on the beautiful idea that gravitation is the result of the curvature of space-time. In other words, Einstein's theory was supported by many *converging* lines of evidence and grounds of credibility.

Another good example of how a theory comes to enjoy acceptance is the Dirac equation, invented by P. A. M. Dirac in 1928 to describe electrons in a way consistent with both special relativity and quantum theory. Dirac was led to the equation primarily by considerations of mathematical beauty. But the equation also resolved a puzzle— namely, that the magnetic moment of the electron was twice as big as previous theory had held it ought to be. The Dirac equation predicted a new phenomenon: the existence of antiparticles. And it shed light on a property of electrons called spin. Eventually it was used to make many very precise experimental predictions that were later confirmed. This example illustrates the fact that, while many factors can lead theorists to entertain a certain hypothesis or build their confidence in it, ultimately it is usually a significant number of precise and correct quantitative predictions that "clinches it." (That, at least, is the case in sciences where controlled and repeatable experiments are possible. However, it is unreasonable to demand the same kinds of confirmation in all fields. For instance, there is a great deal of converging evidence that the evolution of species has taken place, but one cannot predict how particular lineages will evolve. Similarly, one may learn what causes earthquakes without being able to forecast them accurately.)

FROM COPERNICUS TO NEWTON

In a certain sense, one could almost say that Sir Isaac Newton (1643–1727) was the Scientific Revolution. There is much truth in Alexander Pope's famous couplet:

> Nature and Nature's laws lay hid in night.
> God said, "Let Newton be," and all was light.

Newton was a towering peak. There was no one to rival him in physics until the twentieth century. One may think of everything that went before Newton as having set the stage for his great breakthroughs, and everything that came after him—until the twentieth century—as having exploited those breakthroughs.

Three lines of development led to the achievements of Newton: in astronomy, the discovery of Kepler's laws of planetary motion; in physics, Galileo's discovery of the law of falling bodies; and in mathematics, the development of analytic geometry and the use of coordinates by Descartes (1596–1650).

Copernicus, Nicolaus (1473–1543) was born in Torun, Poland, and studied astronomy at the University of Cracow. His uncle, the bishop of Ermland, obtained for him a position as canon of the Cathedral of Frauenburg, an administrative office. Copernicus studied civil and canon law at the University of Bologna and medicine at the University of Padua before obtaining a doctorate in canon law from the University of Ferrara in 1503. He thereupon returned to Ermland, where he acted as advisor to his uncle the bishop and took up his duties as canon. He acquired a wide reputation as an astronomer and was visited in 1539 by Georg Joachim Rheticus, professor of mathematics at the University of Wittenberg, who persuaded Copernicus to publish his ideas on heliocentric astronomy. Copernicus finished his epoch-making work *De revolutionibus orbium coelestium* (*On the Revolutions of the Heavenly Spheres*) shortly before his death, the published copy being presented to him on his deathbed. This is the book that sparked the Scientific Revolution.

In astronomy, the line that led to Newton began with Copernicus (1473–1543), who sparked the Scientific Revolution with his heliocentric theory of planetary motion. It proceeded through the extremely precise observational work of the great Danish astronomer Tycho Brahe (1546–1601). And it culminated in the discovery by Johannes Kepler (1571–1630) of his three great laws of planetary motion (which would have been impossible without Brahe's data). Galileo (1564–1642) was not important in this particular line of development—indeed, he firmly rejected Kepler's crucial idea of elliptical planetary orbits. Rather, Galileo's great contribution to astronomy was the use of telescopes, by means of which he made a series of dramatic discoveries—such as the phases of Venus, the moons of Jupiter, and sunspots—that helped undermine Aristotelianism and the Ptolemaic system. In astronomical *theory*, however, it was Copernicus and Kepler, not Galileo, who made the key advances. On the other hand, in physics Galileo made the important breakthrough, when he discovered the law of falling bodies by applying to terrestrial phenomena the powerful combination of experimentation (carried out with inclined planes and pendulums) and mathematics.

In Kepler's planetary laws and Galileo's law of falling bodies we have examples of precise mathematical laws that apply to specific systems or to a narrow range of phenomena. Today we would call these "empirical relationships" or "phenomenological laws." The genius of Newton enabled

Brahe, Tycho (1546–1601) was born of a Danish noble family. While studying at the University of Copenhagen, his interest in astronomy was piqued by a predicted eclipse that took place in 1560. Studying existing astronomical charts, he found them all to be in disagreement with one another. At age seventeen he decided that "what is needed is a long-term project with the aim of mapping the heavens conducted from a single location over a period of several years," an effort to which (along with alchemy) he devoted his life.

In 1572, Tycho observed the appearance of a "new star" (a supernova). He was able to show that it lay far beyond the atmosphere, contradicting the Aristotelian principle that the celestial realm was unchanging. This impressed the king, who built him an observatory that Tycho named "Uraniborg" (castle of the heavens). Later, Tycho had another, subterranean observatory built nearby named "Stjeleborg" (castle of the stars).

Tycho's were the most accurate astronomical observations ever made (or that *can* be made) with the naked eye. He rejected the heliocentric Copernican theory because he understood that the motion of the earth around the sun would lead to small shifts in the apparent positions of the stars in the sky ("stellar parallax"), and he was unable to observe this. (The stars are so distant that the parallax effect was not seen until 1838.) Thus, he proposed his own geocentric model. He was aided in his last years by Kepler, who succeeded him as "Imperial Mathematicus." The vast wealth of precise data Kepler inherited from Tycho allowed him to discover that the orbit of Mars was an ellipse, not a circle, and to formulate his three great laws of planetary motion.

Tycho was an exotic figure. While a student, he lost part of his nose in a duel and wore a prosthesis made of gold and silver for the rest of his life. At his ancestral castle in Knudstrup, where he entertained on a grand scale, he kept a court jester named Jepp, a dwarf to whom Tycho attributed clairvoyance. Tycho also kept a tame moose, who, after imbibing too much beer one night at dinner, tumbled down a flight of stairs to an ignoble death.

him to see behind these relationships the operation of laws of much greater generality and depth—namely, the law of universal gravitation and the three universal laws of motion. Like all great advances in science, the articulation of these laws led to profound unifications. The first unification was of terrestrial and celestial phenomena. Until Newton, the general and deeply ingrained belief was that the heavens and the earth were wholly disparate realms governed by fundamentally different principles and even composed of different kinds of matter. The "crystalline" heavens appeared eternal, untouched by the kinds of change ("generation and corruption") that characterized the "sublunary" world. It was therefore seen as fitting that the "natural motion" of heavenly bodies should be in perfect circles, for such motion is without beginning or end (as we have seen, even Galileo could not completely free himself from these ancient ideas). What Newton showed, however, is that the very same forces govern both the celestial and the terrestrial realms. The orbits of the planets around the sun, the swinging of pendulums, and the falling of dropped weights all obey the same equations of gravity and mechanics. Indeed, Newton showed that ocean tides could be explained by the gravitational forces exerted by the moon and sun. Interestingly, this had been suggested much earlier by Kepler, but had been ridiculed as superstitious by Galileo.

MATHEMATICS IN A NEW ROLE

The Scientific Revolution, as well as the modern science to which it gave birth, was characterized not only by a happy marriage of mathematics and experiment but also a different way of looking at mathematics and its application to the physical world. A common view at the time of Copernicus and Galileo was that mathematics is *useful* for describing the quantitative aspects of things, but not particularly relevant to *understanding* those things or their causes. For example, the techniques of geometry could be used to predict accurately where heavenly bodies would appear in the sky at particular times, much as a modern train schedule is useful for predicting when trains will arrive at various stations. But just as a train schedule does not tell you what makes the trains go or why, the view of many Aristotelians was that mathematics gave no real insight into phenomena and their underlying physical causes: that was the job of "natural philosophy." (It is significant that scientists were called "philosophers" in Galileo's time, but astronomers were called "mathematicians.")

That is one reason that the heliocentric system of Copernicus did not create much of a stir before the time of Galileo. It was generally seen as merely an alternative method of computation which, while having certain advantages, involved no claim that the earth was *really* in motion. The motion of the earth in the Copernican system was widely understood to be merely "hypothetical," with no more reality than the constructions geometers made to prove their theorems. As long as any computational scheme correctly predicted where things would appear in the sky—"saved the appearances," as they put it—it was regarded as no better and no worse than any other scheme, except in terms of convenience.

Now, as a matter of fact, some of Galileo's telescopic

Kepler, Johannes (1571–1630) was born in Weil der Stadt, Germany. After entering the University of Tübingen to study for the Lutheran ministry, he learned of the ideas of Copernicus and became enchanted with astronomy. He taught mathematics in Graz, but was driven out by the advancing Catholic Counter-Reformation. Finding work with Tycho in Prague, he was eventually forced to leave there for the same reason. (Kepler suffered vexations from his fellow Lutherans, too. They excommunicated him for his views on the Eucharist, and, on another occasion, prosecuted his mother for witchcraft.) Using Tycho's data, Kepler discovered his three great laws of planetary motion. He was enabled to make these discoveries not only by his persistence and mathematical skill, but also by his sound physical intuition, which told him that the sun, as the largest body in the planetary system, somehow exercised a controlling influence on the other bodies. His astronomical thought was deeply influenced by his Pythagorean mysticism as well as Christian theology, which led him to see analogies between the Trinity and the interrelations of the heavenly bodies. It was Kepler who opened the door to the new science, though Copernicus and Tycho had given him the keys. At the end of his book *Harmonices mundi* (*The Harmonies of the World*), in which he announced his third law, Kepler exulted "I thank thee, Lord God our Creator, that thou allowest me to see the beauty in thy work of creation."

discoveries (in particular, the phases of Venus) showed that the Ptolemaic system was no longer able even to "save the appearances," whereas the Copernican system could. That was not enough, however, to prove the earth really moved, for there was on the market an alternative, proposed by Tycho Brahe, to the Copernican and Ptolemaic systems. The system of Tycho saved all the appearances just as well as that of Copernicus, but without having to suppose that the earth moves. (It was therefore embraced by the Jesuit astronomers of the time). In fact, Tycho's system was simply the Copernican system *as viewed from the earth* (or as we might say now, it was the Copernican system as it appears in the "frame of reference" in which the earth is at rest). From a purely "mathematical" point of view, there was no way to decide between the Copernican and Tychonic systems, *and there still isn't,* if one understands the role of mathematics as most did in Galileo's time—that is, if one divorces mathematics from physical causes.

All that changed with Newton, for his laws of motion and his law of gravity provided the critical link between the *mathematical* description of motion and the *physical* causes of motion. Specifically, it related acceleration to force.

To appreciate this point, it may be helpful to consider a simple example. Suppose that a two-hundred-pound man is swinging a little ball around himself in a circle at the end of a long, elastic string. Mathematically, one can just as well consider the ball to be at rest and the man to be in circular motion around the ball. Why is the first description *physically* more sensible (as, indeed, it obviously is)? The reason is that, in the first description, we understand the *forces* that are at work, i.e., the "dynamics." Circular motion involves acceleration, and given the speed of the ball and the radius of its path we can calculate its acceleration. We can also calculate the force exerted on the ball by the string, because we understand strings. (Specifically, we can measure how much the string is stretched, and there is an empirical law, called Hooke's law, that relates the amount a string stretches to the force it exerts.) And what we would find in the sensible frame in which the ball is going round the man is that the force of the string just matches the acceleration of the ball times its mass, satisfying Newton's famous law, $F = ma$.

However, in the second description—where the ball is considered to be at rest, with the man going round the ball—the forces cannot be explained in a physically sensible way. Because the man's mass is so large, the string's force is not enough to constrain him to move in the circle. Therefore, to satisfy Newton's law in the frame of reference where the ball is at rest, large additional forces acting on the man must be assumed. Where do they come from? Nowhere. There is no intelligible physical origin for them; they must be introduced ad hoc. In modern terminology, they are "fictitious forces," and the need to postulate them is a symptom that one is not describing the situation in a *physically* sensible (or "inertial") frame of reference.

Note, then, that given the correct "dynamical laws," one may begin to understand physical reality and physical causes *through* a mathematical analysis. It is only by measuring the stretching of the string, its "elastic coefficient," the mass and speed of the ball, and the radius of its path, and then applying the dynamical laws of Newton to these quantities by means of the requisite *mathematical* calculations, that one arrives at a proper understanding of what is *physically* happening and why. (In the same way, the knowledge of Newton's law of gravity and Newton's laws of motion allows one to see that it really is the earth that is in motion about the much more massive sun.)

This is a mathematization of science far more profound than was understood before Newton, except by a few who had inklings of it, such as Kepler, Galileo, and perhaps Copernicus. When Galileo said that "the great Book of Nature is written in the language of mathematics," he heralded a radically new approach to the physical sciences.

NEWTONIAN PHYSICS

Newton's law, $F = ma$, refers to a single point-like body, which has mass "m" and acceleration "a", and which suffers a force "F". However, it can be applied also to "extended objects" and "continuous media" by conceiving of them as made up of many small (effectively point-like) parts. Consequently, Newtonian mechanics can be used to analyze a vast range of phenomena, including the motion of fluids, the pressure of gases, the flow of heat, the vibrations of sound, and the stresses and strains of elastic solids. As these applications were made during the two centuries after Newton's laws were formulated, an ever greater unification of physics was achieved.

The universal scope and vast explanatory power of Newton's laws, as well as the technique of analyzing things into component parts, had the effect of promoting a "mechanistic" view of the world. As we have seen, the idea that the universe is a machine was commonplace even in medieval thought. However, the medieval thinkers had in mind primarily the clock-like motions of the heavenly bodies. What began to take hold in the 1600s was the idea that *everything* in nature, including plants, animals, and the human body itself, could be understood in mechanical terms.

An important aspect of this mechanistic view was the idea of "determinism." The idea of determinism in physics did not arise from anyone's philosophical prejudices or presuppositions, it came from the mathematics of Newtonian mechanics itself. In Newtonian mechanics, the state of a physical system at a particular instant of time can be completely characterized by a set of numbers, which include both "coordinates" and "momenta." The coordinates specify the positions of the parts of the system at that instant, and the momenta specify their instantaneous velocities. Given all this information at one time (the "initial conditions"), the so-called equations of motion allow one to calculate how these coordinates and momenta will evolve in time. And, generally speaking, this evolution is *unique*. This is different from, say, a game of chess, where from a particular starting position the rules of chess allow a vast number of different games to be played out. In Newtonian mechanics, if one knows the configuration (i.e., all the coordinates and momenta) at one time, then the rest of the "game" is uniquely determined from that point forward (and also backward). That is why, in 1819, the great mathematician and physicist Pierre-Simon Laplace (1749–1827) wrote, "For an intelligence which could know all the forces by which Nature is animated, and the states at some instant of all the objects that compose it, nothing would be uncertain; and the future, as well as the past, would be present to its eyes."

Newtonian physics is also "mechanistic" in the sense of dispensing with "teleology," which played so important a role in Aristotelian science. That is, in Newtonian physics the behavior of a system can be predicted without invoking any "final cause" (any future "end" toward which it is tending,

---

Newton, Sir Isaac (1642–1727) was born in Woolsthorpe, England, and entered Cambridge University in 1661. His genius was soon recognized by Isaac Barrow, Lucasian Professor of Mathematics, who resigned his position in 1669 so that Newton could have it. After his graduation in 1665–66, Newton, because of an outbreak of plague in Cambridge, spent an eighteen-month period in Woolsthorpe, perhaps the most productive period ever in the life of a scientist. It was then that he came up with his theory of colors, discovered the elements of both differential and integral calculus, and had crucial insights that led to his theories of gravity and mechanics.

Because of his development of the reflecting telescope, Newton was elected in 1671 as a Fellow of the Royal Society (the highest scientific honor in England). His publication, soon afterwards, of his discovery that white light is a mixture of all colors of light led to fierce controversy, which made him slow thereafter to publish his results. Nevertheless, at the urging of the astronomer Edmund Halley, Newton published his great work *Philosophiae naturalis principia mathematica (The Mathematical Principles of Natural Philosophy)* in 1684. It is now generally agreed that Newton and Gottfried Wilhelm Leibniz discovered calculus independently of each other and roughly simultaneously; nevertheless, a bitter dispute soon arose over whose discovery was made first, a battle that lasted long after their deaths. Newton's *Opticks* was published in 1704.

or "goal" toward which it is striving). Rather, it is enough to know the *past* state of the system and the laws of physics. This fact contributed to the idea that nature is "blind" and without "purpose." It should be noted, however, that a somewhat more teleological way of looking at Newtonian physics is possible. In the eighteenth and nineteenth centuries, primarily through the work of Pierre-Louis de Maupertuis (1698–1795), Leonhard Euler (1707–83), Jean D'Alembert (1717–83), Joseph-Louis Lagrange (1736–1813), and William Rowan Hamilton (1805–65), powerful ways were developed to reformulate Newtonian mechanics in terms of the so-called "least action principle." A similar principle for optics, called the "least time principle," had been formulated a century earlier by Pierre Fermat (1601–65). The least time principle said that in traveling from some initial point to some final point a beam of light will follow the path that takes the least time. To solve for the light's path using this principle, one must therefore know in advance both where the light begins and where it is going to end up. The analogous principle in mechanics says that *any* system will evolve from its initial configuration to its final configuration by following the sequence of intermediate configurations (called the "trajectory," "path," or "history") that minimizes a quantity called the "action" (usually denoted S).

This way of formulating Newtonian mechanics is mathematically equivalent to the older way of formulating it in terms of forces, in the sense that it gives exactly the same

Laplace, Pierre Simon Marquis de (1749–1827) was born in Beaumont-en-Auge, France. When only eighteen, he so impressed the mathematician Jean d'Alembert with his ability that d'Alembert soon secured for him a position as professor of mathematics at the Ecole Militaire in Paris. By producing in short order a large number of papers on difficult problems in mathematics and mathematical astronomy, Laplace earned a position in the Academy of Sciences at the age of twenty-four. His magnum opus was an immense work, *Mécanique céleste*, whose five volumes appeared between 1799 and 1825. In these books he applied Newton's laws to the enormously difficult task of understanding the motions of the solar system in detail, using sophisticated mathematical techniques developed by himself and others, especially his friend the great mathematician Lagrange. One of his greatest achievements was to prove the stability of the solar system. (Newton had believed that certain instabilities required occasional readjustments by God.) When presented with a copy of *Mécanique céleste*, Napoleon asked why God was never mentioned in it, to which Laplace famously replied: "I had no need of that hypothesis." (When Lagrange heard of this, he is reported to have exclaimed, "Ah! But it is such a beautiful hypothesis; it explains many things.") Laplace also helped to lay the foundations of the theory of probability in his work *Théorie analytique des probabilités*.

Laplace is famous philosophically for his clear formulation of the idea of physical determinism. Politically, he was a survivor, currying favor with whoever was in power. This enabled him to keep his head during the French Revolution and to become, for six weeks under Napoleon, minister of the interior. Napoleon noted in his memoirs that Laplace had been "a worse than mediocre administrator, who searched everywhere for subtleties, and brought into the affairs of government the spirit of the infinitely small."

answers. However, the action-principle formulation is more beautiful, powerful, and profound. In the older formulation, one starts off with as many "equations of motion" as there are coordinates needed to specify the state of the system (and for a complex system that number can be exceedingly large). With the least action principle, however, one starts off with the single fundamental quantity, S, and the requirement that the trajectory minimize it allows one to derive all the equations of motion. One thus sees another kind of unification taking place: many laws (or equations) flow from a single dynamical "principle" involving a single fundamental quantity.

FORCES AND FIELDS

Although gravity is intrinsically the weakest force of nature by far, it is the force that dominates events at astronomical scales of distance and even on terrestrial scales. The reason is that gravitational forces are always attractive, so that the gravitational forces exerted on a terrestrial object (you, for instance) by the vast number of atoms in the earth all add together. By contrast, electromagnetic forces can be attractive or repulsive, and the contributions of the negatively and positively charged particles in matter tend to cancel each other out almost exactly. Thus, electromagnetic forces do not make their presence felt in daily life in an obvious way, as gravity does, even though they actually play the central role in most of the phenomena that we can directly observe. Because electromagnetic forces are more elusive and also mathematically much more complicated than Newtonian gravity, it took a long time and the work of many scientists to unravel their secrets. The most important of these were Charles Augustin de Coulomb (1736–1806), Edward Cavendish (1731–1810), Alessandro Volta (1745–1827), André Marie Ampère (1775–1836), Hans Oersted (1775–1851), Georg Simon Ohm (1789–1854), and Michael

Faraday (1791–1867). The discoveries of these men were ordered, extended, and developed into a unified and coherent theory by James Clerk Maxwell (1831–79), probably the greatest physicist of the nineteenth century.

One of the crucial steps on the way to Maxwell's theory was the idea of "fields." (Interestingly, this enormously important theoretical concept was proposed not by a theorist, but by one of history's great experimentalists, Faraday.) Newton's theory of gravity was based on the idea of "action at a distance." That is, one body exerted a gravitational force directly upon another across the intervening space without any intermediary. Faraday, by contrast, conceived of there being electric and magnetic force fields filling all of space. Electric charges and electric currents produce these fields and also are acted upon by them. One can think of electric fields as being made up of "lines of force" that stretch from positive charges to negative ones and pull them together in a manner not unlike elastic bands.

As it turns out, these fields have lives of their own. They contain energy and act not only upon electrically charged particles of matter but upon each other as well. Indeed, these fields are just as real as material particles. When Maxwell completed his theory, he discovered that its equations implied that waves can propagate in these fields, and that these waves travel at the same speed as light. Indeed, subsequent experiments showed that light actually consists of such electromagnetic waves. Thus, Maxwell's theory achieved a unification of three realms of phenomena that for a long time had been thought to be quite distinct: electricity, magnetism, and optics. In fact, a far larger

> Lavoisier, Antoine (1743–94), "the Father of Chemistry," was born in Paris to a wealthy family. After studying law, his interests turned to science. Although he performed many important experiments, his greatest contribution was to bring order to the theoretical chaos of chemistry. Chemists at that time labored under an upside-down theory according to which substances were thought to burn by releasing something into the air called "phlogiston," rather than by combining with something in the air, namely oxygen. Joseph Priestley, who discovered oxygen, thought it was "dephlogisticated air"; and Henry Cavendish, who discovered hydrogen, thought it was water with extra phlogiston, while oxygen was water lacking phlogiston. Lavoisier showed that combustion was really a process of oxidation, and that the whole idea of phlogiston was mistaken. At a time when many chemists still believed in four fundamental elements—air, earth, fire, and water—Lavoisier made a remarkably accurate table of thirty-three elements (only three of which turned out later to be compounds). He brought rational order also to chemical terminology, which hitherto had been totally confused. (Zinc oxide was called "flowers of zinc"; iron oxide was "astringent Mars saffron"; lead oxide was "red lead" in England and "minium" in France; sulfuric acid was "oil of vitriol," etc.) Lavoisier clarified the distinctions between, and relations among, salts, acids, oxides, and so on, and invented the modern system of chemical nomenclature. During the totalitarian madness of the Terror, charges were trumped up against Lavoisier and he was guillotined. Lagrange observed, "It required only a moment to sever that head, and perhaps a century will not suffice to produce another like it."   •

unification is involved, because electromagnetic forces are responsible for the interactions among and within atoms. Thus, the physical properties of matter (such as heat conductivity, elasticity, opacity, viscosity, and so on), which are studied in "Condensed Matter Physics," as well as the chemical properties of matter, are all based upon the electromagnetic interactions of particles.

While Maxwell's theory involved new forces, phenomena, and concepts, it is at heart a Newtonian theory. Like Newtonian mechanics it is based on a set of equations (indeed, like Newton's, they are "second-order differential equations") that deterministically govern the evolution in time of a set of coordinates and momenta. True, the notion of a coordinate must be broadened to include not only the positions of particles (as considered by Newton) but also the magnitudes of the fields that exist at every place in space. Still, Maxwell's theory involved an extension, not an abandonment, of Newtonian concepts and principles.

Maxwell, James Clerk (1831–79) was born in Edinburgh. He graduated from Trinity College, Cambridge, in 1854, held professorships at Marischal College in Aberdeen and King's College in London, and in 1871 was named the Cavendish Professor of Physics at Cambridge University. While at Aberdeen he wrote a sixty-eight-page prize-winning paper on the nature of Saturn's rings, demonstrating that they could be stable only if made up of disconnected particles. The Astronomer Royal, Sir George Airy, described it as one of the most remarkable applications of mathematics he had ever seen. This led Maxwell to think about the motions of molecules in gases and to apply statistical methods to understanding them. He discovered, independently of Boltzmann, the "Maxwell-Boltzmann distribution" of the velocities of gas particles, and he made other fundamental contributions to statistical mechanics and thermodynamics. Inspired by Faraday's ideas, he next began to work on electricity and magnetism. He developed a theory of the dynamics of "fields" or "lines of force" that reached mathematical completion in four differential equations now called Maxwell's equations, the greatest achievement of nineteenth-century physics. Maxwell was a deeply pious man whose personality was marked by gentleness and modesty; at the same time, he was called "the most genial and amusing of companions." In his last years he selflessly nursed his ailing wife until he was rendered incapable of doing so by the cancer that killed him at the age of forty-eight.

THE TWENTIETH-CENTURY REVOLUTIONS IN PHYSICS

THE THEORY OF RELATIVITY

The idea of the "relativity of motion" long predates Einstein—in fact, it goes back to Newtonian physics. It is the idea that velocity is not the property of one thing, but a relationship between two things. That is, it is meaningless to ask what the velocity of something is; one must ask what its velocity is *with respect to something else*. (A useful analogy is that one cannot meaningfully ask about the angle one line makes, but only about the angle between two lines.) In other words, in Newtonian physics velocity is a "relative" concept. This may seem to contradict what was said earlier about Newton's laws of motion allowing one to determine, by means of an analysis of forces, whether a ball is going around a man or the man around the ball. However, there is no real contradiction: the analysis of forces does not reveal which objects are *moving*, but rather which objects are *accelerating*; and acceleration is an absolute concept in Newtonian physics. (It makes sense to talk about the acceleration of a single object, since acceleration is the velocity of an object at one instant *with respect to itself at another instant*.)

If one is on a train and it suddenly starts to accelerate, one can tell, even if one's eyes are closed, because one feels a force or jolt. However, if the train is not accelerating, one feels no such force, and one cannot tell whether the train is standing still or gliding perfectly smoothly with constant velocity. In fact, it is meaningless to ask whether the train is "really" moving in and of itself; one can only ask whether it is moving with respect to the platform or some other object. In the same way, if there are many objects moving uniformly with respect to one another (none of them accelerating), it makes no sense in Newtonian physics to ask which ones are really moving. What one does in practice is to choose an object on the basis of convenience and measure all motion with respect to it. This is what we meant before by picking a "frame of reference." If one is sitting in the train, it is convenient to measure motion with respect to the train; if one is standing on the platform, it is convenient to measure motion with respect to the platform. However, the "principle of relativity" says that fundamentally it does not matter what frame of reference is chosen.

What does it mean to say "it does not matter"? It certainly matters in terms of how things appear to move. To the man on the train the platform is moving, while to the man on the platform the train is moving. To put it more technically, the coordinates and momenta of objects will be different in different frames of reference. And in Newtonian physics there is a precise rule, called the "Galilean transformation law," that tells one exactly how they are different. (This rule corresponds to our everyday experiences and intuition. For example, if a car goes by you at 40 miles per hour, and another goes by you in the same direction at 75 miles per hour, then to someone in the first car, the second car will appear to be going 35—75 minus 40—miles per hour. That is what the Galilean transformation law says, and it seems to be just common sense.)

Instead, the statement "it does not matter what frame of reference is chosen" means that whatever frame one uses to measure coordinates and momenta, *the coordinates and momenta will obey the same equations.* In other words, the motion of particular objects will look different in different frames, but *the laws of physics will have the same mathematical form in any frame.* That is the key point, and the real essence, of the principle of relativity.

Maxwell's theory of electromagnetism seemed to violate the principle of relativity. It looked as though Maxwell's equations were only true if coordinates and momenta were measured in one special frame of reference. And that would appear to give a natural definition of "absolute velocity," namely velocity *as measured in that special frame.* It seemed, therefore, that there had to be something wrong with either the hallowed principle of relativity or Maxwell's theory of electromagnetism.

This is where Einstein entered the picture in 1905. His aims were actually very conservative: he did not want to abandon either the principle of relativity or Maxwell's theory. And this forced him to take a bold step. He suggested that the Galilean transformation law—the one that seems so commonsensical—is wrong; and that the correct transformation law is the one that had been formulated by a Dutch physicist named Hendrik Lorentz.

Einstein, Albert (1879–1955) was born in Ulm, Germany. Contrary to romantic myth, he excelled in mathematics and physics in school, although, due to a lack of interest, he did less well in subjects that required more memorization. He entered the Swiss Federal Polytechnic School in 1896 and received his diploma in 1901. Unable to land an academic position he began work at the Swiss Patent Office. Einstein had a deep understanding of the physics of his day and the major theoretical issues confronting it, which he had been pondering for years. This bore fruit in Einstein's "miracle year," 1905, when he published three epoch-making papers: his paper proposing the theory of special relativity; his paper on the effect called "Brownian motion" (which showed that atoms are real—something still not universally accepted at that time); and his paper on the "photoelectric effect," wherein he helped lay the foundations of quantum theory by demonstrating that light acts as a particle rather than as a wave in certain situations. From 1909 to 1914 he held professorships in Zurich and Prague. In 1914 he became a professor at the University of Berlin, where he remained until Hitler came to power, at which time he renounced his German citizenship and accepted a position at Princeton University's Institute for Advanced Study. His theory of gravity—the theory of general relativity—was published in 1916. It is one of the great monuments of the human intellect. Einstein knew that he "stood on the shoulders of giants" (as Newton had said of himself). In his study he kept portraits of three men: Newton, Faraday, and Maxwell.

If coordinates and momenta in different frames of reference are related by the Lorentz transformation law, then it turns out that Maxwell's equations work in any frame of reference. Thus, the principle of relativity and Maxwell's theory can be reconciled. However, there is a major catch: *Newton's* laws no longer work in every frame! In other words, Einstein succeeded in saving Maxwell, but at the expense of Newton. Consequently, Newton's laws had to be changed.

What was it about Newton's laws that had to be changed? It was not his three famous laws of motion (including F=ma), nor the laws of conservation of energy and momentum, nor the "principle of least action."

All these things remain true in Einsteinian physics. Really, only one thing had to change, and that was the *geometry of space and time.* The old Galilean transformation law is based on the ideas that three-dimensional space is Euclidean in its properties, and that time is altogether distinct from space. But those ideas turned out to be wrong. The Lorentz transformation laws said—though no one had grasped their real meaning until Einstein—that space and time together make up a four-dimensional manifold that has a different kind of geometry. In this new geometry, even the Pythagorean theorem has to be modified.

We can get some idea of how time is related to space in Einstein's theory by first considering how the three dimensions of space are related to each other. I can choose my three basic space directions (or "axes") to be "forward," "rightward," and "up." (That is to choose a frame of reference in space. It allows me to measure something's position, by saying that it is, for instance, twenty feet in front of me, thirty to the right, and ten above my head.) However, if I turn my body a little to the left, so that I am facing in a different direction than before, the direction I used to call forward, I would now have to describe as *partly* forward and *partly* rightward. In an analogous way, in Einstein's theory the "time direction" of one frame of reference becomes, in another frame of reference, *partly* the time direction and *partly* a space direction. So time and space must be thought of as four basic directions in a single four-dimensional "space-time." In this profound sense Einstein's theory *unified* space and time.

The theory of relativity led to other unifications as well. "Energy" and "mass" turned out to be, in a sense, the same thing (which is the meaning of the famous formula $E = mc^2$). And the three-component electric field and three-component magnetic field of Maxwell's theory turned out to be facets of a single, six-component "electromagnetic" field, rather than distinct entities. In fact, what is purely an electric field in one frame of reference is partly electric and partly magnetic in another frame.

As we have already noted, Newton's theory of gravity also had to be modified. This too involved a new assumption about the structure of space and time, namely that the fabric of four-dimensional space-time is curved. It is this curving or warping that is responsible for all gravitational effects in Einstein's theory of "general relativity." This makes gravity more like electromagnetism, in that gravity is no longer understood to be based on "action at a distance," as Newton said, but on fields. (The "gravitational field" at a given location is the amount that space-time is warped there.) These gravitational fields, like Maxwell's electromagnetic fields, have lives of their own and can have waves propagating in them. The fact that all forces are now understood to come from fields creates the possibility of a "unified field theory" of all forces. Einstein sought such a theory in his later years without success. However, great progress has been made on this problem in recent decades.

HOW "REVOLUTIONARY" WAS RELATIVITY?

In what sense were Einstein's theories of special and general relativity "revolutionary"? They certainly led to conclusions that were profoundly counterintuitive and very surprising. For instance, they showed that it is not absolutely meaningful to say that two events happen at the "same time": it depends on the frame of reference. However, they did not completely overthrow the physics that went before; far from it. Indeed, as we saw, Einstein was led to his theory of special relativity precisely by his effort to *maintain* Maxwell's theory of electromagnetism and the old principle of relativity at the same time. Not surprisingly, therefore, Maxwell's theory was left completely untouched. And the principle of relativity was essentially untouched as well. For example, it is true in Einsteinian physics, as in Newtonian, that velocity is a relative concept while acceleration is an absolute one. Much else in Newtonian physics was also preserved, including Newton's three laws of motion and such basic concepts as force, velocity, acceleration, momentum, mass, and energy (although some of these quantities had to be reinterpreted as vectors in four-dimensional space-time rather than in three-dimensional space).

The word *revolution* is misleading when applied to scientific theories. In a revolution, the old order is swept away. However, in most of the so-called revolutions in physics, the old ideas are not simply thrown overboard, and there is not a radical rupture with the past. A better word than *revolutions* to describe these dramatic advances in science would be *breakthroughs*. In great theoretical breakthroughs, new insights are achieved that are profound, far-reaching, and take scientific understanding to a new level, a higher viewpoint. Nevertheless, many—indeed most—of the old insights retain their validity, although in some cases they are modified or qualified by new insights. Probably the only true revolution in the history of physics was the first one, *the* Scientific Revolution of the seventeenth century. The physics that preceded that revolution, namely the physics of Aristotle, was largely set aside and replaced by something thoroughly different.

We see this in the fact that physics courses in high

school, college, and graduate school do not begin with a study of Aristotelian physics. The details of Aristotelian physics are of interest only to students of history, not to modern scientists as scientists. It is not necessary to know anything at all about Aristotelian physics to do science nowadays. By contrast, before one learns the theory of relativity (or quantum theory), it is still necessary to spend several years studying the physics of the seventeenth through nineteenth centuries. That physics is still profoundly relevant. In fact, many branches of modern physics and engineering still use only pre-relativity and pre-quantum concepts.

Moreover, and this is a crucial point, the Newtonian theory of mechanics and gravity remains as *the one and only correct "limit"* of Einsteinian physics when speeds are small compared to the speed of light, and when gravitational fields are weak. That is, the smaller speeds are and the weaker gravitational fields are, the more accurately do Einstein's answers agree with Newton's answers. The idea of an older theory being the correct "limit" of the theory that replaces it is extremely important. In such a case, the older theory is not strictly speaking right; however, it is not simply wrong either. It would be better to say that it is "right, up to a point."

An example will help make this clearer. Any map of Manhattan, if it's printed on a flat piece of paper, must be wrong, strictly speaking, because the surface of the earth is curved. However, Manhattan is small enough that the earth's sphericity has negligible effects. (It affects the angles on a map of Manhattan by less than one ten-thousandth of a degree.) Indeed, it would make no sense to worry about those effects, because they are dwarfed by other ones, such as the hilliness of Manhattan Island. Therefore, ignoring the earth's sphericity is a reasonable approximation to make if one is talking about sufficiently small areas. And statements based on it are not simply falsehoods; rather, they contain real information and give correct insights into geographical relationships. This is a critical point: an incomplete and inexact description of a situation may be sufficient to convey a completely true insight into that situation. Otherwise, we could never learn anything.

Let us take another example from a "revolution" in physics that hasn't happened yet but is widely anticipated. Both Newtonian and Einsteinian physics are based on the idea that space (or space-time) is a continuous manifold of "points" that lie at definite "distances" from each other. However, quantum theory makes it appear extremely doubtful that one can apply such concepts to the very small. Many physicists expect that our intuitive concepts of space and time will prove to be altogether inadequate for describing anything smaller than a fundamental scale called the "Planck length" (about 10^{-33} cm), and that radically new concepts will have to be used. That is, a new theory will be needed. And it is thought that when this new theory is found, it will show that our intuitive concepts of "space," "time," "point," and "distance" are never applicable to the physical universe except in an approximate sense. That approximation is surely extremely good for distances much greater than the Planck length; nevertheless, it is *always* only an approximation. Supposing all these expectations someday prove to be correct, would it mean that all statements employing the concept of distance (e.g., "the book you are holding in your hands is 8 inches by 5.2 inches by 0.3 inches," or "the distance from my home to my office is 1.75 miles") are false? Obviously not. The concept of distance, while strictly speaking only approximately valid, is such a good approximation in such situations that to cavil at its use would be pedantic and unreasonable. In the same way, one is quite justified in continuing to use Newtonian physics in many situations (including all those that arise in everyday life), even though we know it does not give us an exact and complete account of what is going on.

One final point requires emphasis: Einstein's theory of relativity has nothing whatsoever to do with the foolish idea that "everything is relative." In both Newtonian physics and Einsteinian physics (as in life generally) some things are relative and some things are absolute. For instance, in both Einsteinian physics and Newtonian physics, velocity is relative but acceleration is absolute. In Newtonian physics both temporal distance and spatial distance are absolute, whereas in Einsteinian physics they are both relative, but something called "space-time distance" is absolute. And in Newtonian physics the speed of light in a vacuum is relative, whereas in Einsteinian physics it is absolute (it is the same in every reference frame). The term *relativity* has caused endless mischief. Things are not *more* relative in relativity theory than in Newtonian physics; rather, *different* things are relative and *different* things are absolute.

THE QUANTUM REVOLUTION

Quantum theory was not the brainchild of one man, as was relativity. Many great scientists contributed to its development from 1900 to the mid-1920s, when its basic structure was complete. The major founders of quantum theory include Max Planck (1858–1947), Einstein (1879–1955), Louis deBroglie (1892–1987), Niels Bohr (1885–1962), Arnold Sommerfeld (1868–1951), Max Born (1882–1970), Werner Heisenberg (1901–76), Erwin Schr.dinger (1887–1961), Wolfgang Pauli (1900–1958), and Paul Dirac (1902–84).

Quantum theory has a far better claim to be considered "revolutionary" than does relativity theory. Whereas relativity changed our understanding of space and time, quantum theory fundamentally transformed the basic conceptual framework of all of physics. Physical theories that employ the pre-quantum conceptual framework (whether Newtonian or relativistic) are called "classical" theories.

Even so, it would be misleading to say that quantum theory simply "overthrew" classical physics. Quantum physics is built upon the foundations of classical physics in a profound way. In fact, there is a precise and general procedure for "quantizing" any classical theory, that is, for constructing a quantum version of it. And in the appropriate limit (roughly speaking, when systems are large) the quantum version gives the same answers as the classical version. Moreover, it is not possible to relate quantum theoretical predictions to actual measurements without making use of

---

Heisenberg, Werner (1901–76) was born in Würzburg, Germany. After obtaining his doctorate in physics from the University of Munich in 1923, he worked with Max Born at the University of G.ttingen and Niels Bohr at the University of Copenhagen in the rapidly developing area of quantum physics. At that point, fundamental physics was in disarray, as theorists struggled to find a consistent framework for quantum ideas to replace the existing confused patchwork of insights and methods. In 1925, at the age of twenty-three, Heisenberg discovered this framework and published his "matrix mechanics." (In 1926, Erwin Schrödinger published a "wave mechanics" that was soon shown to be the same theory in different mathematical guise. Heisenberg and Schr.dinger both received the Nobel Prize in Physics.) In 1927, Heisenberg formulated his celebrated and very fundamental "uncertainty principle," which holds that the "coordinates" and "momenta" that classical physics uses to describe the state of a system cannot have definite values at the same time. He continued to make important contributions to nuclear physics, condensed matter physics, and particle physics. During World War II, he led Germany's atomic bomb project. His motives and his commitment to the project have remained the subject of controversy. However, he certainly deplored what he called the "infamies" of the Nazi regime, which he saw as a "flight into insanity that took the form of a political movement."

classical concepts.

An important fact about quantum theory is that it is based on probabilities in a fundamental way. Probabilities are often useful in classical physics, too, but there they are an accommodation to practical limitations. In classical physics, if one had complete information about a system at one time, one could (in principle) know everything about its past and future development exactly, as Laplace noted. There would be no need of probabilities. However, in quantum theory complete information about a system does not uniquely determine its future behavior—only the probabilities of various outcomes. This famous nondeterminism (or "indeterminacy") of quantum theory is obviously of great importance philosophically. Some have argued that it is relevant in some way to the freedom of the human will, an argument that, not surprisingly, is highly controversial.

The probabilistic character of quantum theory leads to very difficult epistemological and ontological questions, which have given rise to a variety of "interpretations." The issues are too complex and subtle to review here. However, it may be of great significance that the traditional interpretation (also called the "Copenhagen," "standard," or "orthodox" interpretation) gives special status to the mind of the "observer" (i.e., the one who knows the outcomes of experiments or observations). The reason for this, in a nutshell, is that *probabilities* have to do with someone's degree of *knowledge* or lack thereof. (If one knows a future outcome, one need not use probabilities to discuss it.) As the eminent physicist Sir Rudolf Peierls (1907–95) put it, "The quantum mechanical description is in terms of knowledge, and knowledge requires *somebody* who knows." Peierls and others, such as the Nobel laureate Eugene Wigner (1902–95), have argued that the traditional interpretation of quantum theory implies that the mind of the observer cannot be completely described in physical terms. If true, this assertion has profound philosophical—in particular, antimaterialist— implications. However, dissatisfaction with the traditional interpretation has led many to embrace alternatives, such as the "many worlds interpretation" or a version of the "hidden variable" or "pilot wave" theories. There is no majority view on these questions among either physicists or philosophers.

None of this philosophical confusion means that quantum theory is in any trouble as a theory of physics. There is no ambiguity or controversy about its testable predictions; and these predictions have been confirmed in countless ways over a period of eighty years, as of this writing. If superstring theory proves to be the ultimate theory of physics, as many leading physicists expect, then quantum theory is probably secure, for superstring theory does not at this point seem to entail any revision of the fundamental postulates of quantum theory.

We have observed that most great advances in physics lead to profound unifications in our understanding of nature. Quantum theory is no exception; it led to one of the most remarkable unifications of all, namely of *matter* and *forces*. In the classical electromagnetic theory of Maxwell, light is made up of waves in a field. However, Planck in 1900 and Einstein in 1905 showed that certain phenomena could not be understood unless light was assumed to come

---

Faraday, Michael (1791–1867) was born in London. He was the son of a blacksmith and his education was, in his own words, "of the most ordinary description, consisting of little more than the rudiments of reading, writing, and arithmetic at a common day school." At thirteen he became an errand boy in a bookshop, and at fourteen he started a seven-year apprenticeship as a bookbinder, which gave him the opportunity to read many scientific books. In 1813, Faraday attended a public lecture by the famous chemist Sir Humphrey Davy, taking copious notes (Davy had pioneered the use of electricity to break apart compounds and had discovered in that way five chemical elements). Faraday applied for a job with him and was at first rebuffed. He soon applied again, sending Davy the notes he had taken. Impressed, Davy hired him as a secretary, then fired him (advising him to go back to bookbinding), then hired him again as a laboratory assistant. Faraday soon became a brilliant experimental chemist in his own right; however, he is most famous for his research in electromagnetism. In particular, he showed that wires moving relative to magnets had electrical currents "induced" in them. (Faraday showed that the same current was induced whether the magnet moved or the wire. This fact was one of the clues that led Einstein to his theory of relativity.) Faraday's law of induction is one of the pillars of Maxwell's theory of electromagnetism. It is even more important that Faraday was the first to articulate the concept of a force "field," an idea fundamental to modern physics. Devout, humble, and generous, Faraday is one of the most appealing personalities among the great scientists.

in discrete chunks, or "quanta," of energy—in other words, particles. These particles of light are now called "photons." The puzzle that something could be both a wave and a particle, a seeming contradiction, was resolved in quantum theory. "Wave-particle duality" was then found to apply across the board. Just as things that were understood classically to be waves were seen to be also particles, so things that were understood classically to be particles were now seen also to be waves—indeed, waves in a field. For instance, the electron is both a particle and a wave in an "electron field" that fills all of space. On the other hand, as Faraday taught us, forces also arise from fields. Thus, both particles of *matter* and the *forces* by which they interact are manifestations of one kind of thing, a field, which is why the basic language of fundamental physics for the last half-century is called quantum field theory. The force between two particles can be understood as being due to "field lines" stretching between them, as Faraday pictured it, or, equivalently, as being due to the exchange of "virtual particles" between them, as Richard Feynman (1918–88) pictured it.

THE ROLE OF SYMMETRY

A crucial role is played in modern physics by the idea of symmetry. In mathematics and physics the word "symmetry" has a precise definition: if a transformation of some object leaves it looking the same as before, then that transformation is said to be "a symmetry" of the object. For instance, rotating a snowflake by an angle of 60° leaves it unchanged, so one says that rotation-by-60° is a symmetry of the snowflake. Altogether, a snowflake has six such "rotational symmetries," since rotating it by 60°, 120°, 180°, 240°, 300°, or 360° leaves it unchanged. To take another example, a soccer-ball pattern has sixty rotational symmetries—i.e., there are sixty ways to rotate the ball that leave it looking the same. A highly developed branch of mathematics called group theory is devoted to the study of symmetry.

Symmetry is found in many of the most beautiful objects of the natural world, including flowers, shells, and crystals. It is also ubiquitous in art, architecture, music, dance, and poetry. Symmetric patterns are found in the rose windows of cathedrals, in colonnades, in friezes, in tile patterns, in arabesques, in French gardens, in the steps of dances, in the arrangements of the dancers themselves, and in the rhyme and metrical schemes of poems, to give but a few of many possible examples. Symmetry has aesthetic power because it contributes to the harmony, balance, and proportion of a thing, and also because it is a *principle of unity*. All the parts of a pattern have to be present in order for symmetry to be realized. Remove one petal of the flower, one point of the snowflake, one column in a colonnade, one rhyme in the sonnet, and the symmetry is spoiled, the unity impaired. As we shall see, symmetry is also a unifying principle in physics. The ever greater unity that we see in the laws of nature is in part the consequence of the ever deeper and more impressive symmetry principles that have been uncovered by theorists.

Symmetry can be possessed not only by physical objects, but also by something as abstract as an equation. In fact, what physicists are most interested in are the symmetries possessed by the very laws of physics themselves. Many of the great advances in physics have entailed the discovery of new fundamental symmetries of these laws. For example, the entire content of the special theory of relativity is that the laws of physics remain exactly the same in mathematical form if one shifts from one "frame of reference" to another by doing a Lorentz transformation. Thus, Einstein's theory amounts to saying that the laws of nature are "Lorentz symmetric." Another example is that Maxwell's equations of electromagnetism have a subtle symmetry called "gauge symmetry," which was discovered by the great mathematician and mathematical physicist Hermann Weyl (1885–1955). Indeed, remarkably, the very existence of electromagnetic forces in nature is a consequence of the gauge symmetry of the laws of physics.

In the twentieth century two other fundamental forces were discovered in addition to gravity and electromagnetism. They are called the "weak force" (or "weak interaction") and the "strong force" (or "strong interaction"). We do not experience them in daily life, since they are only significant at distances smaller than an atom. In the 1960s and early 1970s it was realized that these subatomic forces are also based on symmetries of the "gauge" type, and indeed owe their very existence to the presence of those symmetries in the laws of nature. Gauge symmetries are mathematically quite subtle and far removed from the kinds of symmetries that we can visualize, such as those of snowflakes or flowers. For instance, the symmetry that underlies the strong force is mathematically related to rotations that take place in an abstract space of three "complex" dimensions. (Called "complex" because the three coordinates required to locate something in such a space are not ordinary numbers, but "complex numbers." A complex number is a number of the form $a + ib$, where "i" is the square root of -1.)

In the early 1970s, it was further realized that the three non-gravitational forces (i.e., electromagnetic, weak, and strong) could be understood as parts of a single "grand unified" force. (While this hypothesis has not yet been confirmed definitively, there is much indirect evidence in its favor.) These theories of "grand unification" are based upon even more remarkable gauge symmetries. The symmetry of the simplest such theory, for example, involves rotations in an abstract space of *five* complex dimensions. Besides gauge symmetry, other highly recondite symmetries are suspected to be important in the fundamental laws of physics. One such is called "supersymmetry," which involves in its mathematical formulation the use of so-called Grassmann numbers. These numbers have a peculiar property: if A and B are any two Grassmann numbers, then $A \times B = - B \times A$, rather than $A \times B = B \times A$, as is the case for ordinary numbers.

As a well-known particle physicist has written,

> Symmetries have played an increasingly central role in our understanding of the physical world. From rotational symmetry, physicists went on to formulate ever more abstruse symmetries. . . . Fundamental physicists are sustained by the faith that the ultimate design is suffused with symmetries.
>
> Contemporary physics would not have been possible without symmetries to guide us. . . . Learning from Einstein, physicists impose symmetry and see that a unified conception of the physical world may be possible. They hear symmetries whispered in their ears. As physics moves further away from everyday experience and closer to the mind of the Ultimate Designer, our minds are trained away from their familiar moorings. . . .

The point to appreciate is that contemporary theories, such as grand unification or superstring, have such rich and intricate mathematical structures that physicists must marshal the full force of symmetry to construct them. They cannot be dreamed up out of the blue. Nor can they be constructed by laboriously fitting one experimental fact after another. These theories are dictated by Symmetry.

"THE UNREASONABLE EFFECTIVENESS OF MATHEMATICS"

As science has progressed, the laws of nature have been found to form a unified structure, indeed a magnificent edifice of great subtlety, harmony, and beauty. As this structure is understood more deeply, it is found that ever more abstruse and yet elegant mathematics is required to describe it. Remarkably, much of that mathematics was studied and developed by pure mathematicians solely for its intrinsic interest and beauty long before it was found to be applicable to the natural world. The theory of complex numbers, for instance, was highly developed by the early 1800s even though it had no apparent relevance to science. Yet complex numbers turned out to be necessary to the formulation of quantum theory in the 1920s. Similarly, group theory was developed in the late 1800s and early 1900s, years before it was found to be useful in, and indeed of central importance to, fundamental physics. Many other striking examples could be given.

This history led Eugene Wigner to wonder about what he famously termed "the unreasonable effectiveness of mathematics" in understanding the physical world. There is some deep mystery here. It seems as though the realm of pure mathematics is not something that human beings arbitrarily construct or invent, but a place in which they make discoveries. And many of the most beautiful ideas that they have discovered have proved to be exemplars or patterns for things at the most basic levels of physical reality. The more deeply we look into the heart of nature, the more mathematical it is found to be.

Since 1984, fundamental physicists have become fascinated by a theory of unprecedented mathematical depth called "superstring theory," which many of them suspect may be the long-sought unified theory of all physical phenomena. In this theory, the fundamental constituents of matter are not particles, but loops of string vibrating in a ten-dimensional space-time. Each kind of particle is, as it were, a different note played on this string. One of the greatest physicists of our time, in describing this theory to a science journalist, felt frustrated by his own inability to communicate the grandeur and magnificence of what his research had revealed to him. He said, "I don't think I've succeeded in conveying to you its wonder, incredible consistency, remarkable elegance, and beauty."

The profound mathematical harmony of the world was first glimpsed by Pythagoras in his studies of the vibrations of musical strings. How appropriate it is to think that 2,500 years later the story of physics has returned to vibrating strings! The Pythagorean vision has been vindicated beyond all expectation.

Finally, consider Hermann Weyl's reflection on the mathematical elegance and beauty of science, made in a lecture at Yale University in 1931:

> Many people think that modern science is far removed from God. I find, on the contrary, that it is much more difficult today for the knowing person to approach God from history, from the spiritual side of the world, and from morals; for there we encounter the suffering and evil in the world, which it is difficult to bring into harmony with an all-merciful and all-mighty God. In this domain we have evidently not yet succeeded in raising the veil with which our human nature covers the essence of things. But in our knowledge of physical nature we have penetrated so far that we can obtain a vision of the flawless harmony which is in conformity with sublime reason.

A STUDENT'S GUIDE TO PSYCHOLOGY

INTRODUCTION

Psychology, that "nasty little subject," as William James called it, embraces the full range of actions and events which appear to depend, at least in part, on perceptions, thoughts, feelings, motives, and desires. These very processes, however, also seem to depend, at least in part, on internal biological states as well as external social influences. To complicate matters even further, influences can be "social" only insofar as they are perceived or thought of as such, and can only be "influences" to the extent that they converge on motives, feelings, and desires. These considerations, in turn, reflect or are in some way conditioned by larger cultural and historical influences. In all, then, what James ironically described as a "nasty little subject" is in fact a complex and overarching set of problems and perspectives arising from the abiding project of self-knowledge.

That project, of course, is not *owned* by any discipline or society of scholars or scientists. Self-knowledge includes factors at once biological, genetic, anatomical, medical, social, civic, political, moral, aesthetic—the full range of facts and endeavors that give shape, direction, and definition to a given life. There can be no sharp line establishing just where the psychological domain ends and another begins. Typically, specialists in one domain assume to be more or less settled what those in another accept as a central problem or question. Thus, the political scientist accepts Aristotle's dictum that "man is a social animal," and then proceeds to examine the various forms and foundations of political community. The evolutionary biologist might examine the adaptive advantages conferred by social life; the psychoanalyst, the consequences of withdrawal from the social context.

The history of science leaves no doubt but that such specialized modes of inquiry yield a rich crop of useful facts and basic principles by which to understand a wide range of phenomena. Equally clear, however, is that these gains are not without cost. The principal cost is the narrowing of perspective and the tendency to regard a small part of the overall picture as revealing the essential nature of the whole. The biochemist who accurately summarizes the atomic and molecular composition of the human body has not summarized anything of interest about the human *person* whose body it is. The neurophysiologist who presents an account of the processing of information in the optic nerve does not explain just what makes one scene breathtaking and another prosaic. Needless to say, responsible biochemists and neurophysiologists claim no more than what is warranted by the findings of rigorous research and what seem to be plausible inferences. But the unsuspecting consumer of information, especially when encouraged by the specialist's exaggerated claims, is vulnerable to the "nothing but" fallacy: Optic nerve discharges are essential to normal vision, therefore vision is *nothing but* these discharges!

The sections that follow present findings and theories developed within specialized fields of psychology and kindred disciplines. Reminders are supplied regularly to the effect that these findings and theories form some part of the larger story, but surely not the whole story. Moreover, it can only be the more complete story that allows one to decide finally just how important the various parts of the tale are. Consider, for example, a story such as *King Lear* or *War and Peace*. If these are taken in their wholeness, it becomes clear that the essence of the stories would be unaffected had Lear been a Dutch rather than a British king, or if Pierre had not been quite as tall. The point can be expressed economically: Until one has a defensible and general conception of the essence of human nature, all attempts to establish any one *part* of the story as essential must be premature.

This much granted, it is also clear that what we might come to regard as the "essential" nature of anything must include its composition, its functioning, and the ways in which it manifests itself. It would be odd to discuss the essential nature of human beings, for example, while having no basis whatever on which to classify something as "human"; or to assert that an essential feature of such beings is "thought," but with no clear identification of actions or events as instances of "thought." Though the parts of the story are insufficient, they are nonetheless vitally important if there is to be progress toward the fuller account.

PHILOSOPHICAL PSYCHOLOGY: INVENTING THE SUBJECT

ANCIENT GREECE (AGAIN)

Gnothe se auton ("Know thyself") is the maxim carved into the temple at Delphi honoring Apollo. It is to the world of ancient Greece that one must look for the origins of psychology as a subject of study, a discipline holding out the promise of knowledge. It is to this same world that one turns when considering most of the subjects with which the scientific and scholarly worlds are concerned. Why is this so? The reasons generally offered are many, none of them entirely

---

Pythagoras (c. 580–c. 500 B.C.) was the first to call himself "philosopher." He is credited with dazzling discoveries in mathematics and with the discovery of the mathematical principles of music. His teachings were also influential in shaping some of the main tenets of Platonic thought. Little is known about the life of Pythagoras; he himself, like Socrates, wrote nothing. It appears that he was originally from the island of Samos and later founded an academic society in Kroton (in southern Italy), which was at one and the same time a sort of religious community and scientific school, and which had many devoted followers. He was eventually forced to flee from Kroton to Metapontion, where he later died.

successful. The thesis that explains the cultural achievements of Hellenism as the result of a slave economy permitting the wealthier class to engage in intellectual pursuits proves to be empty. Slave economies before and after the age of classical Greece yield no comparable record, nor are the greatest intellectual accomplishments of ancient Greece those of the aristocracy. Indeed, those who may be said to have founded philosophical inquiry were Greek colonists in Asia Minor, removed from the wealthier centers of the Greek mainland.

A more promising if inevitably incomplete explanation draws upon the nature of ancient Greek religion. The world of the Hellenes was pious, but there was no *official* religion; no set of constraining orthodoxies; neither priests nor gods in possession of "The Truth." What the gods of Olympus enjoyed was an immortality combined with formidable physical powers. None of them was endowed with omniscience, and few of them—and these only occasionally—took any special interest in the humble affairs of mere mortals. On the dilemmas that mark out the subject matter of psychology, philosophy, and the natural sciences, the gods of the ancient Greek world were silent, if not ignorant.

It is said that Pythagoras of Samos (c. 580–c. 500 B.C.) was the first to call himself a *philosopher*—a friend (*philos*) of wisdom (*sophos*). Pythagoras surely was not the first person to raise the questions that still animate philosophical discussions, nor was the Greek-speaking world the first world to come to grips with personal, moral, and social problems. There is ample evidence in the records of older cultures—Mesopotamian, Oriental, Egyptian, Hebrew— of serious thought about such matters. One turns to the world of ancient Greece, therefore, not to locate the date on the calendar when problems of a certain kind first appeared, but to locate the period in which such problems were first subjected to the sustained, critical, and skeptical modes of analysis that continue to characterize philosophy as a distinct subject.

Central to this mode of analysis is the suspension of belief in the authority of tradition, revelation, or the individual. To address a problem in a philosophical manner is to acknowledge that neither revealed truths nor the mere customs of a people, let alone the alleged wisdom or inspiration of some person within the community, can be the last word on any fundamental issue. Understood in these terms,

---

Hippocrates (c. 460–c. 377 B.C.) was the founder of an influential school of medicine situated on the island of Cos, in Greece. Based on direct studies of patients, Hippocrates concluded that the perceptual, motor, and cognitive functions are controlled by the brain (and not the heart, as was commonly believed). Rejecting nonphysical explanations of illness, he was the first physician to accurately describe the symptoms of diseases like epilepsy and pneumonia, and his curative recommendations generally relied on factors that promoted natural healing: fresh air, a good diet, exercise, etc. The followers of Hippocrates wrote more than sixty books on a wide range of medical topics, and his school of medicine required students to take the ethical oath that now bears his name.

philosophy is the foundation on which other and similarly disciplined modes of inquiry—other *disciplines*— were developed, especially the natural sciences.

Before the birth of philosophy as a discipline, one looked to the seer, the oracle, the poet for guidance and for insight into the human condition. The figure most regularly consulted in the Greek world was the blind bard Homer (fl. c. 700 B.C.), whose epic songs, the *Iliad* and the *Odyssey*, record the full range of human passions and powers, and the manner in which their deployment leads to triumph or disaster. Within these works is a skeletal "psychology": Life comes about through the presence of a kind of air (*pneuma*) and soul (*psyche*), with emotion and motivation impelled by events in the heart and chest or throat. There is a "soul" that may leave but return to the body, bearing the contents of a dream; there is a "soul" that, once lost, leaves only death in its wake. As with the great poets and dramatists of every age and culture, Homer provides an account of the human condition that cannot be gleaned either from philosophical or scientific study. But it remains a *poetic* account and, as such, raises a score of questions for each one it seems to settle.

SOCRATES AND PLATO

Thus did the early philosophers of the Greek-speaking world initiate wide-ranging speculation on the nature of reality, of the heavens, of matter itself. But it is with Socrates

(470–399 B.C.) that the focus of philosophical inquiry shifts from inanimate nature to human nature. It is with Socrates that *psychological* problems receive their first clear definition, their first provisional solutions. Broadly classified, these are the *problem of knowledge,* the *problem of conduct,* and the *problem of governance.* Consider first the problem of knowledge: How is it that one can know anything? What is the mark or sign of true knowledge? What means are available by which it can be gained? How can one be sure that one is not simply deluded or hallucinating? Where two incompatible claims about what is "known" come into conflict, how is the difference to be resolved? Is there a difference between knowledge and belief? Knowledge and opinion?

Within the discipline of philosophy, questions of this sort form a special field of inquiry known as *epistemology,* an account or a study (*logos*) of knowledge (*episteme*). These same questions are psychological in that they address such fundamental processes as perception, cognition, learning, memory, judgment, belief. Socrates, it seems, wrote nothing; rather, he staged debates with one or another friend, one or another philosopher, testing the arguments of each and laying down challenges by which to show their errors and inconsistencies. The record of these encounters survives in that most foundational of all philosophical writing, Plato's *Dialogues.* In these, the genius of Socrates and Plato (427–347 B.C.) are merged, as the latter reconstructs and creates whole dramas of inquiry on the broadest and deepest of issues. Although there is no single and entirely stable "solution" to the problem of knowledge, the development or various stages of Socrates' thought on these matters can be reduced to several core principles:

a. Behind the ever-changing and cluttered facts of the perceptible world there is the realm of the unchanging, and it is in this realm that truth is to be found. The evidence for this is provided by mathematics. Whereas any physical or drawn figure with three sides and one right angle can only approximate the true form of a rectilinear triangle, the Pythagorean theorem discloses just this true form. Perceptible triangles come and go; the true form of the triangle is immutable and eternal.

b. In light of (a), the knowledge gleaned by the senses must be tentative, uncertain, changeable. The knowledge worth having is not discovered "out there," but recovered from within the mind's own resources. Pythagoras did not discover the theorem that bears his name by looking at three-sided figures, but by reflecting deeply on the nature of things. The truths of mathematics are truths contained within the "soul" itself, reached by way of philosophical guidance.

What Socrates' thesis asserts, in contemporary terms, is that the comprehension of abstract truths is based on essentially *rational-cognitive* processes rather than those associated with perception and learning. As the abstract truths are not "in" particular things, they are not the result of experience. A knowledge of them, therefore, must be nonexperiential or, put another way, *innate*.

What has been called the *problem of conduct* can be rephrased as a question: How should one live one's life? What is the right sort of life? Nested within such a question are vexing subsidiary questions. Thus: Why should one be good? What is "good"? Why should one's conduct be guided by any consideration other than pleasure or the avoidance of pain? Indeed, is one's conduct guided by any other consideration? Is there any guide to the right life other than custom and the values of one's own culture? Is the right sort of life for one person also right for others, or must each decide on an individual and personal basis? Who or what sets the standard in such matters? When cultural values conflict, how is one to decide which is right? In the realm of values, is anything finally "right" or "wrong," or is it all relative to the culture itself? Within the discipline of philosophy, such questions would give rise to the special fields of *ethics* and *moral philosophy*. But again, these same questions have come to define whole branches of psychological study: emotion and motivation as the sources of behavior; the role of reward and punishment in the control of behavior; the conditions favoring altruism; the social determinants of conduct; the formation of values and their role in life; the sources and nature of interpersonal influence; the nature of moral reasoning and the factors that influence it.

On these matters, too, Socrates' position, as conveyed

in Plato's dialogues, changed somewhat over a course of years, but retained certain key elements that still engage the attention of philosophers and psychologists:

a. Human "psychic" nature is a complex or composite nature: part passionate, part volitional, part rational. Conduct is impelled by the emotions. Conduct also expresses what is willed or desired. Conduct also is answerable to certain judgments rationally reached. Thus, the "soul" is beset by antagonistic, even warring parties, not unlike a charioteer striving to reach a goal but pulled by two radically different horses; one tame and disciplined, the other wild and obstinate.

Plato (427–347 B.C.) is author of those dialogues that established the very terrain of philosophical speculation: the problems of knowledge, of ethics, of politics. His Academy would be rich enough in its offerings to retain Aristotle as a student for nearly twenty years. An Athenian aristocrat, Plato was originally named Aristocles, but his broad shoulders earned him his nickname, which derived from "platon," meaning "broad." After the execution of his teacher Socrates in 399 B.C., Plato visited various Greek cities in Africa and Italy, gaining exposure to the ideas of Pythagoras. In 387 B.C. he returned to Athens and founded the Academy (an institution that remained in operation until A.D. 529, when it was closed by Emperor Justinian). Sometime in the 360s B.C. Plato traveled to Syracuse to tutor Dionysius II, the new king of that city-state, hoping to make of him the sort of philosopher-king he believed to be the ideal ruler. But the effort was a spectacular failure, and Plato returned, with some difficulty, to Athens. There he died in his sleep at around the age of eighty after attending a student's wedding feast.

b. The right life, as with the right actions in a given circumstance, is one in which the ruling authority is that of *reason*, and where both emotion and the will are in the service of reason. The relationship among the three must be harmonious but ordered. Discord at this level is a disease; a species of madness for which only philosophical treatments are likely to succeed.

c. As with knowledge itself, the central rational principles of conduct—the ultimate dictates of right reason— are not visible and "out there," but within the person, needing careful cultivation and refinement. When this is achieved, one emerges as *virtuous*. Cultivation is by way of a rigorously controlled childhood, exposure to exemplary citizens, the removal of corrupting influences. Not many can achieve the desired end, for to do so requires certain native qualities that are essentially genetic in origin.

Finally, the *problem of governance,* as addressed in the dialogues, would begin to lay the foundations for the special field of political science and such subsidiary fields as philosophy of law (jurisprudence), political theory, comparative politics. Although these fields have been richly cultivated within philosophy and political science, they remain largely unattended by psychology, creating a great need for what might be called a *civic psychology*—the psychological aspects of life within an irreducibly civic context.

Plato's *Republic* offers the most developed expression of Socratic thought on the problem of governance, but it is a dialogue that begins not in an attempt to analyze the nature of the good state (*polis*), but the nature of the good man (*anthropos*). The latter needing to be enlarged in order to be seen more clearly, Socrates decides it is best to consider just what it is that makes the *polis* good. This will then offer a model by which to examine the nature of the good man. Note, then, that one of the foundational works in *political* science actually begins as in inquiry in *psychological* science. In keeping with the conclusions reached in the ethical and moral inquiries, the *Republic* defends the notion of a well ordered *polis* ruled by a "philosopher-king" whose just laws regulate the masses in precisely the way reason should shape and direct conduct for the individual. The vice of greed is to be eliminated by the elimination of personal property—even children are to be "owned" in common—and the guardians of the *polis* are to be produced by sound breeding policies (*eugenics*).

ARISTOTLE AND THE NATURALISTIC PERSPECTIVE

Clearly, Plato's dialogues cover nearly every issue that would come to instigate further study and debate in the various fields of philosophy, science, and psychology. Yet, in the form presented in the dialogues, so much is left to conversation, to (mere?) dispute—even to debating tricks of one sort or another—that posterity's debts to Plato are great, but mixed, and somewhat eccentric. There are other debts owed the ancient Greeks, however, especially the physicians of the ancient Greek world who attempted to account for mental and behavioral abnormalities in terms of physical disorders. The students and disciples of Hippocrates (c. 460–c. 377 B.C.) are especially noteworthy. In their commentaries on the Hippocratic approach to medicine they offered suggestive observations of the effects of brain disorders on perception, thought, and movement. They opposed the notion of "sacred" diseases (epilepsy), insisting that every disease should be understood in terms of bodily functions.

By far, however, the most direct debts to the ancient world are owed to Aristotle (384–322 B.C.), a physician's son and the greatest student Plato's Academy would claim. Where Plato's genius expresses itself in dialogues that might even be regarded as theater pieces, Aristotle's treatises are academic, systematic, progressive, detailed—the very qualities that mark out a field of inquiry as a *discipline*. In light of the way the term is correctly used, Aristotle is the first and among the greatest of psychologists; less a *philosopher's* psychologist than a *psychologist's* psychologist. Much of his theoretical work is based on direct observations designed to challenge what is merely speculative. His psychology is never far removed from the natural world in which biological and environmental influences operate. In his major works, the Platonic *psyche*—a disembodied repository of truths, present before the birth of the person and surviving death itself—is transformed into a set of actual powers and processes, tied directly to the life and adaptive potential of a living being.

As Aristotle employs the term, *psyche* is nothing but a principle—a grounding or first principle (*arche*) of living things (*zoon*). In the simplest forms of life the psychic power is expressed in the form of nutritive and reproductive functions. For any creature to be and to remain alive, it must have some means by which to obtain nutrition. For the species itself to continue, the same creature must have the power to procreate. At a level of greater complexity, psychic powers include the power of locomotion. Even plants move

---

Aristotle (384–322 B.C.), a physician's son and sometime teacher of Alexander the Great, was the first and the greatest of ancient Greece's systematic thinkers. His separate treatises and studies were and remain foundational for the traditional scientific and humanistic subjects. Born in Stagirus on the coast of Thrace, at seventeen Aristotle was sent by his guardian, Proxenus (his father had died while he was a young boy), to Athens to complete his education. There he entered Plato's Academy, where he remained for twenty years. After Plato's death Aristotle left Athens and eventually landed at the court of Philip of Macedonia, becoming the tutor to the thirteen-year-old Alexander. On Philip's death he returned to Athens and set up his own school, the Lyceum; here he usually delivered lectures while walking, which led to his followers being labeled the "peripatetics." In Athens his connection with the conquering Alexander was, not surprisingly, a great advantage; but on Alexander's death in 323 B.C., a coup displaced Athens's pro-Macedonian government. Aristotle fled to Chalcis in Euboea, where he died within a year.

in the direction of the sun or, as in the case of the Venus flytrap, to enclose and digest an insect. At yet a greater level of complexity, to these powers of nutrition, procreation, and locomotion is added that of *sensation*, which, according to Aristotle, is central to the very definition of *animal*. An entity qualifies as animal insofar as it is capable of sensation. At still greater levels of complexity there is a power Aristotle refers to as intellectual, referring here to the ability to learn and remember based on experience. The animal kingdom displays all varieties of such ability. With the mature and healthy human being, however, there is a power added to all these that is special in its own right: the power or faculty of *reason* (*nous*), which is distinguishable from intelligence. By way of intelligence, a creature can learn and recall specifics. Reason, in the form of what Aristotle at one place calls *epistemonikon*, allows one to comprehend and to frame general and universal propositions. This is the power of genuinely abstract thought, which is the grounding not only of mathematics and logic but also of the rule of law. It is the power that allows human beings to give and to understand *reasons* for acting, and thus to be held responsible in light of abstract moral and juridical precepts. This focus on rational power is not, however, at the expense of perceptual sources of knowledge. In important respects, Aristotle is a commonsense psychologist, not skeptical about the functions of the senses, not dismissive of any power or process that is widely distributed within a given species.

The theory of human nature advanced by Aristotle is often referred to as *hylomorphic*, the roots here referring to the ancient Greek words for matter (*hule*) and form (*morphe*). The "soul," Aristotle argues, is "the form of the body." To understand the sense in which "form" is intended, it is useful to consider Aristotle's own example: If the soul were an eye, vision would be its form. Thus, the hand of a statue is "a hand in name only," for it does not perform the functions for which hands are intended or designed. Thus understood, a living thing is what it is *essentially* owing to a defining form of life or activity. It is in this sense that man is a rational animal; it is in this sense that something materially looking like a man but lacking all rational powers would be a man "in name only."

Aristotle developed a systematic psychology that included the biological, the social, and the political dimensions of life in such a way as to provide a full-scale theory of human nature. Within the general theory are special subsidiary theories of learning and memory, motivation and emotion, cognition and abstract thought; subsidiary theories anticipating developments in genetic psychology, sociobiology, gender studies. At the center of his psychological speculations is a more general *teleological* theory: The regular occurrences or features of the natural world are what they are for a purpose. Nature does not traffic in accidents but in lawful relationships. To understand any natural process, then, is to comprehend not merely the physical causes at work but that "final cause," the *that for the sake of which* the various links in the chain are formed. To identify man as a rational animal is to raise the fundamental question as to just what the rational powers are for; the end for which they were intended. Rationality here has the same status as, for example, the wings of a bird. The latter are present *for the sake of flight*. And rationality is present for the sake of . . . what? Aristotle's answer is that rational powers are employed in the service of securing a flourishing form of life, a life of happiness (*eudaimonia*), but where happiness is understood not as the sensuous pleasure of animals and children but the deep and enduring pleasure of a life rationally lived. Just how one goes about this task is a measure of that person's virtue or moral excellence (*arête*) and the basis on which that person answers to a certain *type*. Indeed, Aristotle's ethical writings, and his classic treatise *Rhetoric*, provide the foundations for theories of personality.

There is scarcely a topic that would come to give psychology greater definition over the course of centuries that was not addressed by Aristotle and for which Aristotle is not among the cited pioneers. To the extent that the subject itself is reducible to a definition, it would be the works of Aristotle that would yield the essential terms of the definition: *Psychology is the study of those perceptual, cognitive, and/or rational powers and processes by which organisms enter into relations with their physical and social environments in order to achieve ends determined by their specific natures.*

PHILOSOPHICAL PSYCHOLOGY AFTER ARISTOTLE

The development of psychology from its founding in ancient Greece to the seventeenth century is chiefly the work of (1) philosophers employing the method of introspection, who, consulting the nature of their own thoughts and actions, advanced general theories about the nature of mind and mental life; and (2) medical practitioners recording the effects of disease and injury on such psychological processes as perception, thought, memory, and movement. These centuries not only hosted some of the most inventive and agile minds of intellectual history, but also saw the founding of the modern university. Especially worthy of mention here are Saint Augustine (354–430), whose *Confessions* is a veritable treatise on psychological processes; Saint Thomas Aquinas (1225–1274), who remains one of the most discerning of the commentators on Aristotle and whose own origi-nal thought touched on the subjects of perception, cognition, and free will; and William of Ockham (1280–1349), who wrote at length on the cognitive basis of the concept of universals. By the twelfth century, the once modest abbey schools had been expanded to offer instruction in a variety of subjects, the school at Paris coming to have authentic status as a university featuring a curriculum in what would now be called the liberal arts.

From the middle of the seventeenth century, however, the possibility of a distinct *mental science* emerges as a realistic project. Three of the most influential figures in this development are René Descartes (1596–1650), Thomas Hobbes (1588–1679) and John Locke (1632–1704). The three are celebrated representatives of the great age of science that was the seventeenth century, the century of Kepler, Newton, Boyle, Galileo, Wren, Huygens. Descartes made significant contributions to mathematics and optics, earning the title of father of analytic geometry. Hobbes, after visiting the great Galileo and mastering the newly discovered laws and principles of mechanics, wrote a foundational treatise on human nature based on just such mechanistic principles. Locke, a fellow of the Royal Society and medical doctor, composed one of the most influential treatises in the history of philosophical psychology, *An Essay Concerning Human Understanding* (1690). It was an explicit attempt to develop a mental science along the lines of Newtonian science. On Locke's account, ideas are composites of more elementary sensations, held together as a result of associational forces. The mind is completely "furnished" by experience, its elementary sensations akin to Newtonian corpuscles, its associational laws functioning as a kind of gravitational force.

Although Descartes is known famously for a *dualistic* theory, according to which mind is immaterial and utterly unlike any property of physical bodies, his psychological theories of perception, emotion, and motivation are utterly physiological at the level of explanation. Only the abstract rational powers of man are excluded from this scheme. Locke, though somewhat noncommittal, is clearly inclined to accept physiological processes as the grounding of all psychological processes, and Hobbes defends a radically *materialistic* psychology without reservation. In this productive age of philosophical psychology, therefore, the most influential works are those that mirror developments in the physical sciences and seek to pattern a scientific psychology along the same lines.

It is in the following century, that eighteenth century of "Enlightenment," that Newtonian science reaches nearly

---

Descartes, René (1596–1650), was one of the geniuses of his age, a major contributor to mathematics, optics, and the biological sciences. His *Discourse on Method* and his *Meditations* established the "Cartesian" position on the nature of philosophical inquiry and the primacy of reason. Born near Tours, France, and educated at a Jesuit school, Descartes, like many of his contemporaries, came to believe in a strict separation of reason and faith. After spending much of his early life in Paris, in 1629 he went into seclusion in Holland for twenty years, jealously protecting his privacy by often changing residences. In 1649 he left Holland at the invitation of Queen Christina of Sweden to become her philosophy tutor. The uncongenial climate and early hours forced upon him by the queen led to his death soon after he arrived. Despite the theologically problematic implications of his philosophical work, Descartes never abandoned his Catholicism. At the time of his death his fame was such that many believed him to be a saint; by the time his body arrived back in France, relic-gatherers had been so many, and so enthusiastic, that his remains were considerably lighter.

religious levels of discipleship. Advances in physics and especially applied physics encourage and defend the conviction that no problem, no matter how complex, is immune to the explanatory range of science. "Social engineering" surfaces in the great works of the period, sometimes in the form of pamphlets and treatises, more powerfully in the form of revolutionary rhetoric and upheaval. The politics of the Enlightenment is largely a political psychology that would justify one mode of governance and reject others on the grounds of a "correct" theory of human nature. With Locke, David Hume, and the other British empiricists, it is the *authority of experience* rather than tradition or scripture that must settle matters of fact and matters of principle. In the patrimony of Descartes, writers in the Enlightenment adopt an official skepticism before the claims of history, rejecting all that cannot be vindicated in the arena of systematic observation and rational analysis. Needless to say, life is more complicated than this, and the various worlds of social and political life would (repeatedly) prove this Enlightenment "gospel" to be too thin by half.

It is in the same century that progress in medicine and more particularly in what would now be called neurology adds measurably to the thick book of clinical findings on the relationship between mind and body, brain and thought. By the end of the century Franz Gall (1758–1828), the father of *phrenology*, offers any number of compelling anatomical observations leading to the conclusion that specific psychological processes depend on specific regions of the cerebral cortex. The decades following these claims are devoted to testing them experimentally in order to establish where and to what extent specific psychological functions are located within the brain. Thus is the systematic study of *localization of function* launched, and thus does it continue to the present time.

PSYCHOLOGY AS SCIENCE

There is no sharply distinguishable period of time when psychology left its philosophical moorings and set sail independently. In point of fact, the very nature of the subject makes "independence" more a slogan than a reality. The development of a science, physics included, is based on any number of suppositions and orientations which are neither validated by the science nor contained within its own resources. Basic questions must be settled first (or seem to be settled), and only then can the inquiry begin. If, for example, physics is taken to be the study of matter and energy and the laws governing their behavior, then there must be some basis on which to defend this definition—some basis on which to identify a relationship as a *law*, an entity as *matter*, an influence as *energy*, etc. In a word, every special field of scientific inquiry has ineliminably *metaphysical* foundations, where "metaphysical" refers to the interrelated philosophical issues of *ontology* (the study or consideration of just *what there is* or what has *real being*) and *epistemology* (the critical examination of knowledge-claims and the means by which knowledge is acquired). There can be no science that is "independent" of these foundations, for it would be a gate swinging without hinges. There is, alas, a continuing blindness or resistance to this truism displayed by the vast majority of contemporary psychologists. There is, however, a different sense of "independent," and one that was invoked by many scientists in the nineteenth century eager to free their subjects from what they took to be merely philosophical

    Mill, John Stuart (1806–1873), is generally considered to be the leading philosopher of the nineteenth century in the English-speaking world. His defenses of an empiricistic alternative to rationalism are authoritative and of enduring influence. In his logical works he developed formal principles of experimental science that continue to guide research strategies. The eldest son of philosopher James Mill, John Stuart's upbringing was, to say the least, intellectually rigorous; he knew Greek at age three, for example, and as a teenager, greatly influenced by the writings of Jeremy Bentham, he formed his own "utilitarian society." The extreme rigidity of his father's rearing methods probably caused Mill's mental breakdown, marked by severe depression, at age twenty-one. After recovering, his career took off. From 1823 to 1858, Mill worked for the British East India Company, rising from clerk to chief of the examiner's office. In 1851 he married the widow Mrs. Harriet Taylor, who had long been one of his intellectual companions, sometimes co-authoring articles with him. Mill served as a Member of Parliament from 1865 to 1868, winning the seat despite his refusal to spend any of his own money on the election (on the grounds that doing so would amount to buying his seat). Before his death in Avignon, Mill managed to write an enormous number of articles on a broad array of topics, the contents of which helped him to earn a reputation as a political progressive.

squabbles. The most influential defenders of this sense of independence included Hermann von Helmholtz (1821–1894) in the German-speaking world, and John Stuart Mill (1806–1873) in England. Each in a different way set down the principles of experimental science, adapting these to the study of mental processes. Helmholtz grounded his theories in the physiological sciences. Mill resisted this, insisting that an experimental science concerned with the "laws of mind" could be prosecuted in a manner distinct from studies of the "laws of body." Both, however, were influential in creating a climate that nurtured the growth of a scientific and *experimental* psychology.

Wundt, Wilhelm (1832–1920), is something of the "father" of modern experimental psychology, for it was Wundt who established the first academic laboratory devoted to psychological research (Leipzig, 1878–79), and the first journal in which such research could be published and widely distributed. His students were chosen by many leading universities eager to establish psychology departments on the "German" model. The son of a Lutheran clergyman, Wundt was born in the village of Nekarau, Germany. A physiologist by training, and a former assistant to Hermann von Helmholtz, Wundt taught the first academic course in psychology in 1862 at the University of Heidelberg. Much of Wundt's work focused on the senses. However, he was also concerned with identifying the "structure" or fundamental elements of consciousness through careful attention to conscious experience. Structuralism, as this mentalist approach to psychology is called, would later come to be rejected by American functionalists and behaviorists.

Among the pioneers of the new science were Wilhelm Wundt (1832–1920) at Leipzig and William James (1842–1910) at Harvard. Each established a laboratory within his university in which to study basic perceptual and mental processes. Priority is usually given to Wundt, whose Leipzig laboratory was established in the period 1878–1879, though James would note in his writings that his Harvard laboratory was up and running as early as 1875. But Wundt deserves priority on a basis more important than mere chro-

---

James, William (1842–1910), was America's greatest philosopher and most incisive psychologist. His *Principles of Psychology* remains a landmark in the field. In 1875 James introduced experimental psychology at Harvard University, establishing what was probably the first academic psychology laboratory in the United States. The brother of novelist Henry James, William was born in New York City in 1842. After abandoning an early love for painting, James entered the Harvard School of Medicine in 1864. He received his M.D. in 1869 but at the same time was overcome by a bout of severe depression. Finally in 1872 he began to teach physiology courses to undergraduates at Harvard, eventually moving on to teach courses in psychology and philosophy. In 1898 he first identified himself in print as a pragmatist, the school of thought with which he is now most often associated. But his interests ranged widely, and he was unique among his contemporary colleagues in the emerging field of psychology for his genuine openness to questions concerning religion and the supernatural. This latter interest led to an invitation to deliver the Gifford lectures at the University of Edinburgh in 1901. These lectures were published the next year as *Varieties of Religious Experience,* which remains a classic. He died of heart failure in 1910.

nology. He founded a journal in which psychological findings could be published. He established graduate programs of study that would confer doctoral degrees on those who would then establish psychology programs at any number of universities in Europe, Great Britain, and the United States. Wundt in these respects is the modern "father" of experimental psychology, though William James's *Principles of Psychology* (1890) would come to be its most developed expression. The character of this new science would be shaped by developments within the psychological domain but also and more importantly by two developments in more or less distinct domains. One of these was specific, the other by way of an accumulation. The specific development was Charles Darwin's (1809–1882) immensely influential works; the other, the work of many hands over a succession of decades. It is instructive to examine briefly how each of these was incorporated into the emerging discipline of psychology.

DARWINIAN EVOLUTIONARY THEORY

Darwin's *Origin of Species* (1859), *Descent of Man* (1870) and *Expression of the Emotions in Animals and Man* (1871) came to dominate thinking in psychology nearly from the time of their initial publication. In defending a continuity theory of mental development, according to which human mental powers are mirrored throughout the animal kingdom, though to a lesser or altered degree, Darwin gave impetus and credibility to the study of the adaptive behavior of animals. Research programs addressed to instinctual behavior, mating and sexual selection, developmental processes from infancy to adulthood, species comparisons, studies of exceptional types—the full panoply—were to spring up seemingly overnight. Social institutions and practices now were to be understood in terms of selection pressures and challenges to survival. Racial comparisons were now framed in terms of relative degrees of evolution, with predictable and comforting racist explanations of socio-economic strata. Mental illness was now understood in the language of adaptation. The long held thesis of *essential-*

Darwin, Charles (1809–1882), studied for careers in medicine and theology at the University of Edinburgh and Cambridge University, respectfully, but neither could maintain his interest. Rather, he committed himself early to the study of natural phenomena, earning the (unpaid) position of naturalist on the HMS *Beagle*, a ship headed for an exploratory trip along the Pacific coast of South America (1831–36). Its voyages supplied Darwin with the data that would eventually be incorporated into his landmark *Origin of Species*, published in 1859. In the intervening years Darwin married, fathered nine children, made a name for himself in the scientific community as a naturalist, wrote a popular account of his travels on the *Beagle*, and began reading the work of Thomas Malthus. *The Origin of Species* was concerned almost exclusively with evolution in the nonhuman world, but in *The Descent of Man* (1871) Darwin applied his theory of evolution by natural selection to man.

ism, according to which a thing is what it is owing to an unchanging and essential aspect of its nature, now gave way to *contextualism,* according to which things adapt to the conditions under which they must struggle for survival. The overarching perspective was one that favored a form of psychological theory known as *functionalism:* The task of psychology is to establish the function of various psychological states and processes in the task of survival and successful adaptation. Instead of asking what is the essence of mind, the right question becomes, *What is the function of mental events?* It would become a central principal in William James's psychology that the one unfailing mark of the mental is action directed toward an end. James's student, E. L. Thorndike (1874–1949), expressed the same idea in the form of his famous *Law of Effect:* Behavior is more or less likely depending on the effects it produces. What functions to secure a satisfying state of affairs becomes ever more dominant; what leads to pain and suffering, ever less frequent.

ADVANCES IN NEUROPHYSIOLOGY AND NEUROLOGY

By 1800 there was evidence that muscles were activated by an electrical force, a theory at first contested but finally established early in the nineteenth century. Between 1810 (Sir Charles Bell) and 1822 (Francois Magendie) anatomical research revealed that the sensory and the motor functions of the nervous system were anatomically distinct. Information reaches the brain by way of sensory nerves entering the spinal cord on the dorsal surface; motor commands from the brain to the muscles exit from the spinal cord on the ventral surface.

Following the lead put in place by Gall—and his critics—research in the nineteenth century left no doubt whatever but that specific regions of the cerebral cortex, the cerebellum, the medulla, and deeper levels of the brain served specific functions. Paul Broca (1824–1880) discovered a lesion in the left frontal lobe in postmortem studies of the brain of his aphasic patient. *Broca's aphasia* left little doubt but that the crowning achievement of human psychological development—language—was a brain-based capacity. By the 1830s a coherent theory of reflex function had been developed by Marshall Hall and others. In the 1840s Helmholtz and others produced experimental data showing that the speed of nervous conduction—authoritatively regarded as almost infinitely rapid just a decade earlier—was a rather sluggish twenty to forty meters per second. Not long thereafter Emil Du Bois-Reymond (1818–1896) obtained evidence in support of the view that the electrical events in the nervous system were chemically created by a process now recognized as ionic. Step by step, the neurology clinic and the neurophysiology laboratory established ever firmer support for an essentially *physiological* psychology and for the "mental" being little more than a code word for what were finally chemical and physical processes. As would be repeated during the seasons of behavioristic psychology, the position here is characteristic of the "nothing but" fallacy mentioned in the introduction: The mind (life, love, virtue, etc.) should be understood as "nothing but" chemical, physical processes. A form of village credulity—never in short supply—is needed to adopt such a stance.

BEHAVIORISM

One abiding goal within the scientific community, at least since the nineteenth century, has been to demonstrate the ultimate unity of science. It was the aim of Ernst Mach (1838–1916) at the close of that century, and it was the aim of his intellectual descendants, the *logical positivists* of the 1930s. Although substantial disagreements can be found among the major figures in the movement, there was (and is) general agreement on these key points: First, that science is a distinct form of inquiry, its claims finally having to be settled at the level of relevant measurements and observations; second, that the subject matter of science is at least in principle *observable*, either directly or by way of observable effects; third, that the ultimate "stuff" of reality is *physical*—there is no ghostly stuff, mental or otherwise; finally, physics is not "metaphysics." No special place need be reserved for divine purposes or hidden designs. The first product of this perspective in psychology was the school of *behaviorism*, defended with great rhetorical flourishes in the works of John B. Watson. It received oblique support from the pioneering research on conditioned reflexes by Ivan Pavlov. And it was brought to its greatest conceptual maturity and influence by B. F. Skinner.

Evolutionary theory explains variation and stability in the characteristics of living things in terms of purely natural processes. Members of a given species display natural variations, some of these better equipping certain members to meet challenges to survival. In time, the naturally selected characteristics become more frequently represented within the breeding pool, the successful "types" becoming the common type. In the most general terms, behaviorism

---

Pavlov, Ivan (1849–1936), the Nobel Prize–winning Russian physiologist, established the principles of classical conditioning. These principles grounded the philosophical theory of "associations" in the actual physiology of the nervous system. Pavlovian psychology was a form of behavioristic psychology, to be replaced by the quite different "Skinnerian" version. Born in central Russia in the village of Ryazan, and the son of an Orthodox priest, Pavlov was originally intended for the priesthood, but, like Darwin—indeed, partly because of Darwin—he decided to pursue a scientific career instead. At the University of St. Petersburg he studied chemistry and physiology, receiving his doctorate in 1879. Pavlov did not set out to contribute to psychology; the experiments that made him famous began as studies of digestion, which was the topic of research for which he won the Nobel Prize in 1904, and he was dubious about the new discipline of psychiatry. A sometime outspoken critic of the Soviet government after the October Revolution, his worldwide fame and usefulness to the Communist Party as an exemplar of its scientific progressivism kept him from persecution until his death at the age of eighty-seven.

is predicated on the assumption that the same processes are at work at the level of adaptive behavior. Those responses that result in positive consequences become a more common feature of the organism's repertoire. Unsuccessful and maladaptive behavior is "extinguished."

It was Ivan Pavlov (1849–1936), the Russian physiologist who had won the Nobel Prize for research on digestion, who established the procedures by which to "condition" certain behavior to specific environmental stimuli. His research established that a basic biological reflex—salivation in response to food being placed in the mouth—could be brought under the control of a previously ineffective stimulus such as the ringing of a bell. The sequence, BELL-FOOD, repeated frequently would result in salivation to the sound of the bell, food no longer being required. This is the well-known paradigm of *classical (Pavlovian) conditioning*. Pavlov also showed that, once conditioned, a response such as salivation was elicited by a range of stimuli falling along the same continuum as that used to establish the conditioned response. Thus, once salivation is conditioned to a tone of 5000 Hz., the response will also be elicited by tones of 1000, 2000, 6000, etc. Under these conditions, Pavlov observed that the amount of salivation was progressively diminished as the test-stimuli became progressively less similar to the initial ("conditioned") stimulus. This common effect illustrates what is called *stimulus generalization*. With differential conditioning—

for example, where during conditioning food is paired with a given tone but where other tones are presented without food—the conditioned response is more sharply "tuned" to the value of the conditioned stimulus, illustrating the process of *stimulus discrimination*.

Pavlov theorized that these behavioral effects reflected processes occurring in the cerebral cortex of the animal.

Watson, John B. (1878–1958), was the "father" of modern behaviorism. His lectures and writing were directed toward a relentless defense of an objective science of psychology, one that would take observable behavior as the discipline's sole subject matter. His "Psychology as the Behaviorist Views it," which appeared in *Psychological Review* in 1913, was a rallying call to an entire generation of psychologists. Born near Greenville, South Carolina, Watson received his B.A. and M.A. degrees from Furman University. At twenty-two, he entered graduate school at the University of Chicago, where he was strongly influenced by the pragmatist-functionalist ideas of Chicago professors John Dewey, George Herbert Mead, and James Rowland Angell, and where he began to develop his behaviorist theory. In 1908 he moved to Johns Hopkins University as a full professor, and five years later launched his attack on mainstream psychology. A man equipped with the instincts and drive of a missionary, Watson published in 1919 a popular introductory textbook that played a large role in advancing his theory. Like his disciple B. F. Skinner, he hoped his work would spur widespread social reorganization. In 1920 he was forced to resign from Johns Hopkins after his wife and former graduate student initiated divorce proceedings. He spent the rest of his career as an advertising executive in New York.

His was a radically physiological theory of conditioning. In America, John B. Watson (1878–1958) launched a relentless defense of an essentially behavioristic psychology, which downplayed the specific physiological processes proposed by Pavlov but otherwise took Pavlov's findings as supportive of a purely behavioristic school of psychology. Behaviorism is committed to the study of observable behavior, making no assumptions about the operation or even the existence of a "mind" or "mental" events on which the behavior allegedly depends. The rationale is that the subject matter of science is what is observable. One can observe the behavior of others, not their minds. With few exceptions, behaviorism is opposed to *mentalism*, this term referring to theories that explain behavior as the result of mental events and processes.

Owing to the influential writing of B. F. Skinner (1904–1990), behaviorism also was long opposed to explanations of adaptive behavior based on physiological processes or events in the brain. It was a central precept within behaviorism, as B. F. Skinner would have behaviorism understood, that a purely descriptive science of behavior was under no obligation to locate the internal processes or mechanisms associated with adaptive behavior. The facts of behavior are just *there* to be observed; their factual standing is unaltered by observations made at some other (e.g., physiological, genetic, anatomical) level of observation. In keeping with this perspective, behavioristic psychologists locate the factors shaping or controlling behavior in the environment external to the organism rather than in that inner environment of such interest to biologists. Hence, critics of behaviorism often labeled it the psychology of "the empty organism."

Skinner, B. F. (1904–1990), was, with the exception of Sigmund Freud, the most influential psychologist of the twentieth century. His studies of conditioning and learning in rats and pigeons came to be veritable models of the most complex human activities, including wagers and gambling, child rearing, interpersonal relations, war itself! After majoring in literature at Hamilton College, Skinner moved to New York City to pursue a writing career. This having failed, he removed to Harvard University, where he took his Ph.D. in 1931. In 1936 he took up academic residence at the University of Minnesota, and there, building on the work of John Watson, began to develop his influential, and often controversial, behaviorist theory. His novel, *Walden Two* (1948), depicted the creation of a utopian society based on behaviorist principles. Skinner was not afraid to subject his own family members to such principles: his second daughter spent much of her infancy in his "baby box," a device that allowed him to strictly control environmental influences on her development. After a brief stint at Indiana University, Skinner ended up at Harvard in 1948, where he remained for the rest of his career. In 1971 he published *Beyond Freedom and Dignity*, a volume of political and social thought whose title speaks volumes about the implications of Skinnerian psychology.

NEUROPSYCHOLOGY AND COGNITIVE NEUROSCIENCE

A behavioral science need not be indifferent to internal states and processes. Largely through the very refinements in behavioral measurement and control achieved by behavioristic psychologists, it became possible, from the 1950s and thereafter, to examine the physiological and biochemical substrates of highly specific aspects of perception, learning, motivation, emotion, and social interaction. As of the new millennium, it can be said that the most active and expanding field in psychology is that of *cognitive neuroscience,* which comprises specialists in experimental psychology, computer science, neurology and neurophysiology, artificial intelligence, and philosophy of mind.

One of the foundations of cognitive neuroscience is the specialty of *neuropsychology.* Over the decades, the subject of this specialty has been subsumed under different headings: physiological psychology, psychobiology, biological psychology, medical psychology. Sharp distinctions cannot be made here. In general terms, however, both neuropsychology and medical psychology have been concerned chiefly with the functions of the human nervous system in health and disease. Physiological psychology, psychobiology, and biological psychology are less specifically concerned with human beings and more broadly devoted to studies of nonhuman animals. What research and theory in all these areas—under all of these headings—have in common are relationships between psychological processes or capacities or abilities and specific neurophysiological, neurochemical, and neuroanatomical substrates.

As noted above, observations of this sort appear as early as the Hippocratic school of medicine in ancient Greece. Only after centuries of study, often interrupted by silent periods lasting still more centuries, were accurate anatomies of human and nonhuman nervous systems discovered and recorded. Not only did details await the discovery of the microscope in the seventeenth century, but the actual mode of function of the nervous system awaited the discovery of instruments by which to record electrical events. Only in the twentieth century was there firm proof of the existence of structurally independent neurons, the cellular units of the nervous system. Only in the twentieth century was there firm proof that the mode of transmission of information from one neuron to another is by way of chemical *transmitters*.

The twentieth century began with anatomical evidence of neuronal specialization; it ended with computational and visualization techniques capable of displaying in real time the actual activity in various structures throughout the entire brain. Combined with systematic studies in clinical neurology, the developed "brain sciences" now are able to identify, often at the cellular level, those processes and events reliably associated with the full range of perceptual, cognitive, behavioral, motivational and emotional aspects of human and nonhuman animal psychology. That science of phrenology that had been launched by Gall at the end of the eighteenth century has evolved in ways perhaps only he would have imagined at the time.

The emergence of cognitive neuroscience has been gradual. Its foundations were laid by basic research in sensory and perceptual processes: studies of sensory threshold, attention, information processing, target recognition, short-term memory, etc. Decades of basic research in these fields led to the modeling of mental functions as a set of *modules* of this general type:

1. Impinging stimuli initiate responses in the sensory organs.

2. The sensory response electrically "codes" such features of the stimulus as intensity, size, location.

3. The coded signals are stored briefly where they can be compared with previously stored information.

4. Comparisons and weighting determine whether the new information is retained or erased.

5. Retained information is given weights as a result of those emotional or motivational processes with which they are associated.

6. Outputs from memory stores initiate processes associated with behavior.

7. The moment-to-moment consequences of behavior are reported back to the system by way of feedback circuits, thus providing something of a closed-loop ("servomechanistic") device capable of adapting to changing conditions.

This is a simple and clearly incomplete sketch of some of the "modules" presumed to operate as animals and persons face one or another environmental challenge. Theories based on this sort of model seek support from studies of the effects of brain lesions, selective stimulation and recording from specific brain sites, and postmortem studies of patients known to have suffered from one or another psychological deficit.

Many leading figures in the field of cognitive neuroscience have adopted theories of this sort. Complex psychological processes are assumed to arise from the activation of more elementary processes. "Problem-solving" on this account is a term that actually covers a range of distinguishable functions: selective attention, sensory coding, perceptual registration, memory, motivational and emotional "gain" or amplification, recruitment of instinctual or acquired behavioral adjustments. These separate functions are thought by many to depend on various functional *modules* within the brain, interconnected by specific pathways and integrated in different ways to solve different classes of problems.

Despite the enthusiasm with which such theories have

been advanced and defended, it should be noted that there is no clear understanding of just what it is about an anatomical structure—a collection of cells—that makes it a "module," nor is it at all clear that calling such cells a "memory module" is even intelligible. It is the *person* or *animal* who remembers the left and right turns in the maze. This much is clear. However, to claim that some structure in the brain "remembers" is, to say no more, surely less than clear. The rationale or motive behind such expressions arises from the "unity of sciences" perspective which, in its more popular form, seeks *reductionistic* explanations. The aim is to explain complex events by reducing them to a set of simpler or more elementary events, in much the way that the properties of water might be reduced to those of hydrogen and oxygen in the appropriate combination. The value of this approach, its defenders would contend, is that it replaces subjective and ambiguous accounts with explanations based on objective scientific principles. There are, however, a number of compelling reasons for resisting reductionism:

1. Attempts to reduce psychological phenomena to physiological events are likely to result not in the explanation but in the elimination of the very phenomena of interest. For example, to reduce the experience of color to discharge patterns in neurons and brain cells is to capture nothing that gives the experience its felt and immediately known quality.

2. There is a somewhat self-defeating feature in reductionistic schemes. After all, what makes events in the ner-

vous system of interest to psychologists is the reliability with which these events are associated with, alas, *psychological* events and processes. To seek to eliminate these is to seek to eliminate just what it is that gives the nervous system its importance within psychology.

    3. The very concept of an *explanation* is understood in radically different ways in various contexts. The victim of a fatal shooting died (a) owing to wounds and loss of blood, (b) because he was known by his assailant as in possession of a large amount of money, (c) because the money in question had been stolen from the assailant and (d) because the assailant was being blackmailed by the mob. There is no basis on which to declare (a) the best explanation simply because it is based on "objective scientific principles."

    4. There are still good reasons for assuming that the very logic of scientific inquiry and explanation, as it has been developed in physics, is finally inapplicable to the range of events (cultural, aesthetic, moral, judicial, political, interpersonal) that constitute the very content of psychology. Thus, reductionistic strategies are not merely premature but finally misguided.

FREUD AND DEPTH PSYCHOLOGY

It is often overlooked that the widespread influence of "Freudian" psychology arose from the work of a clinical neurologist, practicing in Vienna at the close of the nineteenth century, and having as his main objective a clearer understanding of the causes of certain "hysterical" symptoms. Sigmund Freud (1856–1939), like Darwin, did not intend to change the world of thought. What he discovered in his clinical practice was a relationship between certain symptoms (paralyses, blindness, severe anxiety) and the *repression* of psychologically disturbing thoughts or past experiences. In the manner of the conservation laws in physics (which had been discovered in Freud's lifetime), there seemed to be a *psychic* energy that was also "conserved," but able to express itself in various ways. Thus, the hysterical symptom could be regarded as the *physical* manifestation of a process of *psychic* repression.

By 1900 Freud would borrow liberally from the evolutionary theory that was now generally accepted by the scientific community. Darwin offered strong arguments in support of the claim that the psychological aspects of man fell along the same continuum as that containing the mental life of nonhuman animals. Freud accepted the notion that survival-instincts impel human actions in a manner widely displayed in the animal kingdom. Nature equips animals to find pleasure in what finally determines the survival of the individual animal and its species. The suckling infant is motivated not by considerations of nutrition but by sheer sensual pleasure. Thus did Freud offer a theory of *psychosexual development,* marking out the stages, from infancy to adulthood, of dominant patterns of sexual activ-

---

Freud, Sigmund (1856–1939), began his professional career as a practicing neurologist. His encounter with the "hysterical" symptoms of neurotic patients led him to study the use of hypnosis as a diagnostic and therapeutic approach. He was then led to that theory of repression that is foundational for the entire "Freudian" psychology of mental illness. Born in Freiberg, Moravia, Freud's family moved to Vienna when he was four; he would remain in that city until forced to leave by the Nazis in 1937. His volatile and unhappy marriage commenced in 1886, the same year in which he set up his private practice in the treatment of psychiatric problems. In formulating his theories Freud was heavily influenced by the work of French neurologist Jean Charcot and his colleague in Vienna, Josef Breuer (with whom, as with almost all his colleagues and students, he would later have a falling out). Freud did not gain wide public attention until 1908–09, when the first International Psychoanalytic Congress was held and he traveled to the United States to deliver a series of lectures. The father of six, including daughter Anna, who later also became a well-known psychoanalyst, Freud died of cancer in England in 1939.

ity. The culminating stage is that of *heterosexual procreative sexuality*, the means by which the species is preserved.

The central theoretical element in Freudian psychology is that of *unconscious motivation*. The reasons persons give for their actions may well be impelled by unconscious desires that would lead to social ostracism were they publicized. In the course of *socialization*, the developing child, naturally inclined toward self-gratification and otherwise obedient to the *pleasure principle,* comes face-to-face with that *reality principle* imposed by adult society. The conflict between these competing principles is lifelong. Total domination by the reality principle results in a less than authentic life; total yielding to the pleasure principle renders one unfit for life within society.

Because the sources of desire and action are (allegedly) unconscious, the person is unable to understand at the deeper and most informing levels just why his or her aims and frustrations, triumphs and failures take the turn they do. The psychological disturbances that plague the adult are, according to the theory, first planted in childhood, surviving now as what Freud called a "childhood remnant." It is only through a process of *depth analysis* that the patient can be returned to the very time when the seeds of the conflict were sewn. It is the analyst, not the patient, who finds the deeper meaning and significance in the patient's anxieties, hopes, failures.

The Freudian perspective is so broad in its expression as to be by now resistant to clear definition. It is a perspective that has influenced philosophy, literary criticism, law, the arts, and yes, even psychology! It may be that psychology is now less influenced than the others. As for the influences, perhaps the most general are these:

1. A skepticism toward rational explanations of individual actions and cultural values. On the psychoanalytic understanding as Freud developed it, rational accounts typically are *rationalizations,* designed to put in a favorable light initiatives and practices actually driven by animalistic motives of the Darwinian stripe. Darwin + Freud = (much of) evolutionary psychology and sociobiology.

2. A relativism in the matter of basic moral precepts. "Morality" on the psychoanalytic account is a set of concessions that the pleasure principle must make to the reality principle if unacceptable degrees of censure and punishment are to be avoided. There are no absolutes in the moral sphere, reached by reason and valid in all contexts; only the daily collision between ego and world, primal urges and their repression, primal urges and their symbolic transformations.

3. A species of psychological determinism. The fixity of the stages of psychosexual development, the iron nature of evolutionary forces and laws, and the requirements imposed by civilization all work to render the individual life one of discontent, conflict, quiet desperation. This is all "in the cards," so to speak, with amelioration in the form of a resigned understanding achieved through depth analysis.

It was Freud's own expectation, variously affirmed and questioned, that the psychoanalytic terms and principles would ultimately be recast as facts within a developed science of neurophysiology. His ultimate objective was the biologizing of the mind, with psychoanalytic theory being but a first step. However, the theory itself is essentially a kind of *narrative* with little by way of the testable, the measurable—in a word, the scientifically knowable. It is something of a story about human nature; something of a gothic novel which many find to be eerie, and many others just *true*—as in "true to my life."

So much ink has been spilled in defense and in defiance of Freud's writings that little would be gained by adding to either side of the ledger here. The actual practice of psychotherapy, as reported by therapists themselves, is only loosely indebted to Freud and derives nothing from the theory by way of a "how to" guide to diagnosis and treatment. The Darwinian elements are neither more nor less compelling in Freudian thought than in Darwin's own or, for that matter, in current books and articles that treat these elements as gospel. The recognition that should guide students confronting Freudian theory for the first time is that, as a "theory," it scarcely measures up to what one demands of scientific theories; moreover, that it is in the very nature of Freudian "theory" to be adaptable to nearly any evidence that might be arrayed against it. Perhaps it is best understood as a richly imaginative account of human nature, as one might find in especially probing works of fiction.

THE SOCIAL CONTEXT

If one considers the three dominant "deterministic" perspectives in psychology—the behavioristic, the neuropsychological, and the genetic—it becomes clear that each and all must find a daunting challenge in many of the findings in social psychology. Studies of bystander-effects, of mock prisons, of peer pressure, of obedience, all point to the power of the immediate context on behavior and judgment. The famous and controversial research of Stanley Milgram (1933–1984) required volunteers to deliver what they had reason to believe were painful and potentially lethal shocks to human subjects in experiments (falsely) described as studies of the role of punishment in learning and memory. Approximately two-thirds of the subjects ("teachers") in the study continued to increase what they were told was the voltage of shocks delivered to "learners" who were actually collaborators in the research. As the latter feigned pain and suffering, no more than a bit of encouragement from laboratory assistants was needed to keep the "teachers" obedient to the aims of the research. Comparable effects were obtained by Philip Zimbardo (b. 1933) in studies that simulated the conditions of imprisonment. Randomly assigned to the category of "prisoner" or "guard," Stanford students participated in research that called upon some of them, in their assigned capacity of uniformed guards, to maintain order among the prisoners. In a matter of days the student-guards displayed sadistic conduct toward now clearly submissive "prisoners," many of whom were willing to betray their peers for an extra blanket or sign of approval.

What is profoundly suggestive about such research is the degree to which a veritable lifetime of reinforcement for decent behavior is seemingly undone by contextual factors involving pressures toward conformity. But profoundly missing in discussions of such findings are reflections on that reliable minority that resists such pressures and retains a fidelity to worthy principles. It is not too much to propose that the truly interesting subjects in research of this sort are those more or less ignored, as attention is heaped on the ever-pliant majority.

Experiments in social psychology often reflect a highly realistic quality generally missing in laboratory investigations in psychology. The subjects in a social psychology study are often placed in realistic settings or are asked how they would respond in such settings. The settings, often contrived with remarkable inventiveness, have permitted studies of prejudice, self-appraisal, the strong inclination toward consistency, vulnerability to external presssures, etc. Over the decades of such research yet another conception of human nature has been crafted; one in which the sources or causes of significant actions may well be different from the explanations the actual actors give. Although not psychoanalytic in nature, this conception of human nature again focuses on determinants of action and perception falling beyond the conscious control or awareness of persons. Of course, this conception is based on trends violated at least some of the time by some subjects. More-over, it is often the case that the very framing of the context or the social dilemma restricts participants to a far greater extent than one would find in the real world. The importance of the research and theory is that it provides vivid reminders of the conditions that might disarm persons and find them doing things which, on reflection and in a composed state, they would avoid.

HUMAN DEVELOPMENT: MORAL AND CIVIC

Imaginative and systematic research conducted in the past half-century leaves no doubt but that infants enter the world with highly developed perceptual capacities and more than the rudiments of cognitive, problem-solving processes. It is also clear that, from the earliest stages of vocalization, the very young child is, as it were, *speaking*— which is to say emitting vocalizations according to basic rules—and that it is this language that adult language comes to replace. The child is, indeed, a "little linguist."

Controversy still surrounds the question of those nurturing conditions that are essential to normal psychological development. Laboratory research with nonhuman animals, for all the attention paid to it, has actually settled none of this. Perhaps the most influential program of such research was that conducted by Harry F. Harlow (1905– 1981), who deprived infant rhesus monkeys of access to their mothers, allowing them only to cuddle up to cloth or wire dolls. Harlow's findings indicated that severe maladjustment was a reliable consequence of such deprivation, partly reduced in those animals having at least a doll to attach themselves to.

Needless to say, no such research was required to inform intelligent persons that children removed from all human contact from the earliest stages of life would more than likely acquire maladaptive modes of behavior. But a knowledge of the destructive consequences of extreme deprivation offers no information whatever as to the essential conditions for a normal and flourishing course of development. The thick literature of biography and autobiography is sufficient to establish that no fixed formula is available here. Persons have failed miserably in life even after a privileged childhood of love and care; others have achieved utterly successful and rewarding lives after a childhood of misery and abuse. Thus, there is no iron law establishing the dos and don'ts of childrearing. There is, however, a common sense fortified by a knowledge of the ingredients often found in the developmental histories of those whose lives are worthy of emulation and respect.

It was a fixture in the philosophical and psychological treatises of the ancient Greek and Roman worlds that children are to be reared for *citizenship*, and that this required systematic schooling in what is often translated as *virtue*. The Greek *arête* translates readily as *moral excellence* and refers to a set of behavioral and emotional dispositions, powers of self-control, and the adoption of worthy goals. At the peak of its civilization, Rome instituted methods of education that included in the curriculum dress codes, posture, tone of voice, rhetorical skills, and, of course, the "martial arts." Clearly, the manner in which children are reared for civic life will reflect the form of civic life adopted by the larger culture. Rome did not exist for its citizens; they lived for the glory of Rome. In the Western democracies, based on respect for the dignity of the individual and committed to preserving the basic liberties of the individual, a different regimen or curriculum is required; but there is now not great agreement on what that should include. Nor is there much activity within psychology devoted to the question.

To the extent that there is a *psychology of moral development,* it is partitioned into a small part of the field of *cognitive psychology* and a comparably modest portion of the field of *social psychology.* With cognitive psychology, there is a long but spotty history of interest in how children come to comprehend moral problems and the basis on which they reach solutions. The early work was done by Jean Piaget (1896–1980), who employed the method of storytelling. Thus: A man's wife is deathly ill and needs medication. The nearest pharmacy has the medicine but is charging a price vastly greater than the husband can afford, refusing to give it to him unless he pays. The husband waits till the pharmacy closes for the day, then breaks in, steals the medicine, and returns home with it to save his wife's life. Did the husband do the right thing? If so, why? If not, why not? If he is caught, should he be punished? Etc.

Piaget discovered that the very young child solves all such problems in terms of the consequences of an action:

If it is punished, it is wrong; if rewarded, it is right. Older children and young adults judge the morality of actions not in terms of how others react to them but in terms of principles personally adopted. Piaget distinguished the two cognitive periods as *heteronomous* (Greek *hetero* = other; *nomos* = law) and *autonomous* (Greek *auto* = self); those whose moral laws or rules are imposed by others are said to be in the heteronomous period, and those who are self-ruling are said to be in the autonomous period. Building on Piaget's work years later, Lawrence Kohlberg (1927–1987) studied a wide range of age groups in various cultures and advanced a multi-state theory of moral development. It is a precept within Kohlberg's theory that the stages must be passed through successively and that, though the stages are related to age, age alone is no guarantee of refined moral judgment.

Many social psychology studies, as noted above, have considered the conditions under which persons behave altruistically; the conditions that incline persons to deceive others; the conditions under which even basic principles seem vulnerable to peer pressure or the demand characteristics of the "situation." Whereas the Piagetian and Kohlbergian approaches consider cognitive processes generating moral judgments, social psychologists tend to consider the social aims persons seek to achieve by judging or acting in certain ways. Thus, whether or not one comes to the aid of others in distress depends to some extent on whether anyone else is present and, if present, whether that other person is inclined to get involved. There is, it seems, a difference between how in cognitive terms one *judges* when and whether altruism is called for, and if, in the actual situation, one *acts* altruistically. Beyond these generalities, psychology at present offers little guidance or even a plan of research on such a central issue as civic development.

ABIDING ISSUES: AN EPILOGUE

As should be clear, psychology as a discipline of study and research might well include the full range of human endeavors at the level of the individual person, the social collective, and even national groups. It cannot be a fault of psychology that it is unable to shed bright light on all such endeavors. Less than two centuries old as an independent field of experimental science, psychology continues to refine its methods and, to a lesser extent, its overall perspective on just what topics are to be central to its mission. At present, the overall perspective is still method-bound. That is, psychology as it appears in the major texts and major academic departments is committed to methods of inquiry that, by their nature, dictate the problems to be addressed. This is a backward arrangement. Rather, certain problems and issues should be chosen as in some sense "right" for the discipline, followed by the development of methods suited to just these problems and issues.

Committed to laboratory study and statistical modes of description and analysis, psychology continues to have trouble with actual individual persons, individual minds setting out to achieve goals that are personal, arising from motives that may be quirky, and guided by considerations that are various and shifting. Rather, the approved methods offer fairly reliable statistical descriptions of data drawn from collectives and pooled in such a way as to wash out individuality itself. In striving to be a science of *everyone*, it is unable to say much of consequence about *anyone*.

Then, too, the three domains in which the psychological side of human life expresses itself most vividly—the civic (political), the aesthetic, and the abstract—have been grossly neglected, presumably because the official "methodology" is unable to adapt itself to such matters. There is no political psychology as such; just a rag-tag collection of ad hoc studies on, e.g., voting behavior. There is no aesthetic psychology as such; just some work on what persons perceive and appreciate in, e.g., paintings or music. And, for all the talk of a "cognitive revolution," there is little within the discipline that explores the relationship between abstract thought and, e.g., principles of justice and fairness.

There is much to be done!

打开一扇窗户，看看大学的专业

教授带你"逛"专业

由 130 多位浙江大学的老师和 30 多位学生一起完成，给将要成为大学生的学生们的书。

采用问答的方式，通过"身在专业"的名师大家和同学的生动经历，引领大家"逛"专业，帮助年轻人打开大学的窗，了解认识大学的专业。

希望这本好看、好懂、好用的"专业指南"，能在你选择未来发展方向时，助上一臂之力。

步入学科堂奥的台阶，通向开放未来的知识桥梁

人文社会科学基础文献选读丛书

浙江大学相关学科具有较高学术水平和丰富教学经验的教授和博导，几经筛选、精心校译，帮助读者面对浩繁的文献和有限的时间，以最短的时间、最少的阅读量、最可靠的方式，准确地掌握相关学科最重要的内容。

第一辑（已出）：

《西方哲学基础文献选读》　包利民　编选
《历史学基础文献选读》　　包伟民　编选
《社会学基础文献选读》　　冯　钢　编选
《经济学基础文献选读》　　罗卫东　编选
《政治学基础文献选读》　　郎友兴　编选
《管理学基础文献选读》　　张　钢　编选

第二辑：

《文艺学基础文献选读》（已出）　徐　岱　沈语冰　编选
《语言学基础文献选读》（已出）　施　旭　编选
《传播学基础文献选读》（即出）　李　岩　编选
《心理学基础文献选读》（即出）　王重鸣　编选
《法学基础文献选读》（即出）　　夏立安　陈林林　编选
《教育学基础文献选读》（即出）　徐小洲　编选
《宗教学基础文献选读》（即出）　王晓朝　编选

图书在版编目（CIP）数据

自然科学·心理学：汉英对照 /（美）巴尔,（美）罗宾逊著；刘慧梅，潘寅儿译. —杭州：浙江大学出版社，2015.5

（学科入门指南）

书名原文：A student's guide to natural science; a student's guide to psychology

ISBN 978-7-308-14260-1

Ⅰ.①自… Ⅱ.①巴… ②罗… ③刘… ④潘… Ⅲ.①自然科学—指南—汉、英②心理学—指南—汉、英 Ⅳ.① N-62 ② B84-62

中国版本图书馆 CIP 数据核字（2014）第 303656 号

浙江省版权局著作权合同 登记图字：11-2014-340 号

学科入门指南：自然科学·心理学

（美）史蒂芬·M.巴尔 （美）丹尼尔·N.罗宾逊 著
刘慧梅 潘寅儿 译

策划编辑	葛玉丹
责任编辑	陈佩钰（yukin_chen@zju.edu.cn）
封面设计	项梦怡
出版发行	浙江大学出版社
	（杭州市天目山路 148 号 邮政编码 310007）
	（网址：http://www.zjupress.com）
排　版	杭州立飞图文制作有限公司
印　刷	杭州杭新印务有限公司
开　本	889mm×1194mm 1/32
印　张	8.25
字　数	220 千
版 印 次	2015 年 5 月第 1 版　2015 年 5 月第 1 次印刷
书　号	ISBN 978-7-308-14260-1
定　价	28.00 元

版权所有　翻印必究　印装差错　负责调换

浙江大学出版社发行部邮购电话（0571）88925591